Notes on Computational Fluid Dynamics:
General Principles

AF698852

# Notes on Computational Fluid Dynamics: General Principles

**Christopher J. Greenshields**

and

**Henry G. Weller**

**CFD Direct Limited**
https://cfd.direct

CFD Direct Limited
Reading, RG4 7AN, UK.
`https://cfd.direct`

First published April 2022

Copyright © 2022 CFD Direct Limited

The rights of C.J. Greenshields and H.G. Weller to be identified as authors of this work have been asserted by them in accordance with the UK Copyright, Designs and Patents Act 1988.

All rights reserved. No part of this publication may be reproduced, stored in a retrieval system, or transmitted in any form or by any means, electronic, mechanical, photocopying, recording or otherwise, without the written permission of the publisher, or as expressly permitted by law.

All trademarks used herein are the property of their respective owners. With the exception of "CFD Direct" (a registered trademark of the publisher), the use of any trademark in this text does not vest in the author or publisher any trademark ownership rights in such trademarks, nor does the use of such trademarks imply any affiliation with or endorsement of this book by such owners.

ISBN 978-1-3999-2078-0 (paperback)

British Cataloguing-in-Publication Data
A catalogue record for this book is available at the British Library.

Typeset by C.J. Greenshields in 11pt MLModern using LaTeX.
Illustrations and cover design by C.J Greenshields using Inkscape.

# Contents

# Preface

This book is written for people who use, or are planning to use, computational fluid dynamics (CFD) in their work, research, study or recreation time.

Its purpose is to provide the *knowledge* to help the reader build their confidence to undertake CFD analysis, repeatedly, to a defined standard, in a timely manner[†]. It forms part of our mission to *make CFD accessible and inclusive*, alongside our work to manage, maintain and develop the free, open source software OpenFOAM on behalf of The OpenFOAM Foundation Ltd at `https://openfoam.org`.

The book taps into our experience in CFD, in particular with OpenFOAM. At the time of publication in 2022, it is 33 years since Henry Weller (HW) wrote the first lines of code of the "Field Operation and Manipulation" (FOAM) software which became OpenFOAM in 2004.

Since 2008 Chris Greenshields (CG) has provided CFD training with OpenFOAM, delivering over 300 courses (over 650 days) to 3,000 participants around the world. The training continues today at CFD Direct Ltd (`https://cfd.direct/training`), teaching repeatable *procedures* and *workflows* and the use of OpenFOAM and supporting software to deliver reliable CFD solutions.

---

[†]from Robin Hoyle's definition of 'competence', in *Complete training: from recruitment to retirement*, 2013, p69.

This book does not replace training, but provides the supporting knowledge of the physics and numerical methods that underpin CFD. The user of any CFD software should benefit from reading this book, but they can only relate its contents back to the software if the code is open source, as it is with OpenFOAM.

CFD is challenging because it combines five sub-disciplines which are complex in their own right: (1) creating a computational mesh for the problem geometry; (2) fluid dynamics and physical modelling; (3) numerical methods; (4) data processing; and, (5) computers and programming.

This book concentrates mainly on fluid dynamics, modelling and numerical methods, finishing with a some examples that briefly describe some mesh generation and data processing. It presents:

- general equations of fluid dynamics that are the basis of a CFD solution;
- common physical models, including turbulence modelling;
- the finite volume method to generate equations in linearised matrix form;
- methods to compute solutions of the matrix equations;
- algorithms used to couple systems of equations;
- the practical application and numerical implementation of common boundary conditions;
- some sample problems.

We named the book *Notes on Computational Fluid Dynamics* because we present the subject matter as a set of short topics, or *notes*, spanning exactly two pages. This means each topic appears in its entirety on the left and right pages of the opened book. We use this format to break down the material into small pieces that are easier for the reader to digest.

The book presents CFD from the perspective of an engineer (or scientist), rather than a mathematician. CFD is an application of science so requires the qualities of an engineer to design, analyse, build and test, in order to create and run CFD simulations and interpret the results.

The finite volume method is the principal numerical method used in CFD. It is conceptually an engineering method that uses a computational mesh to define control volumes for physical processes. It is presented as such here, rather than as a mathematical construct which views the mesh as regions which specify distributions of variables.

Of course, we must use mathematics to describe physics, models and methods, but we do so in a way which is more familiar to engineers. Mathematics is just another form of language which we combine with verbal language in a manner that we hope conveys sufficient meaning while avoiding both ambiguity and unnecessary complexity.

For example, we describe a vector as a physical property, such as force and velocity, which has magnitude and direction, illustrated using an arrow. We do not describe it as a "three-dimensional vector space over the field of real numbers", using mathematical notation $\in \mathbb{R}^3$.

We avoid the excessive use of sub- and superscripts within our mathematical notation. For example, when we describe an increase in time $t$ by an increment $\Delta t$, we write $t := t + \Delta t$, where ":=" means that the right hand side is evaluated and assigned to the variable on the left. This avoids the need to distinguish the "old" and "new" $t$ using an index $i$ by $t^{i+1} = t^i + \Delta t$.

Numerical methods are ultimately programmed into a computer, so it is better that the mathematical notation represents typical programming language. The assignment notation "t :=", for example, clearly indicates a change to $t$ already stored in computer memory.

We have tried to reduce the need for the reader to search back through the book for information, to make it easier to follow. One way we hope to achieve this is by reminding the reader what the mathematical symbols represent, *e.g.* specific internal energy $e$ is defined in Sec. 2.15, but the reader is reminded what $e$ represents in Sec. 2.17 and Sec. 6.13.

When a topic refers back to something previously discussed, we have included references to direct the reader to the relevant section, page or equation number. To help the reader find something quickly, we have provided a fairly extensive index and list of symbols.

The book is designed to be self-contained so that the reader does not need to search through other texts to find some missing information. We have therefore dispensed with a long reference list at the end of the book in favour of footnotes within the pages themselves. Footnotes clarify minor points and present the origins of the science that is being used in CFD.

We use diagrams, graphs and images liberally to illustrate the topics we present. Each 2-page topic typically includes 1-2 figures, alongside equations. Some thematic figures are reused to reinforce certain concepts, redrawn with labelling relevant to the particular topic.

Each chapter includes one or more summary topics to give an overview of the main concepts presented within it.

Music has always been playing in the background during our

work in CFD. As a light-hearted distraction, we chose to begin each chapter with relevant musical lyrics, which we think are just a poetic as anything found in literature.

The book is shaped by the people that taught, encouraged and supported us, past and present. CG would like to thank: Pat Leevers, Colin Wood and Bill Knox, for teaching him the value of writing, using diagrams and consistency; Richard Jones for many stimulating discussions about effective teaching; Steve Camp for his infectious enthusiasm and insightful questions; and, Akshai Runchal, Nils Basse and Bill Jones for their correspondence on some important topics.

HW would like to thank Chris Marooney, Bill Jones and David Gosman for their teaching and guidance in CFD.

Thanks to our colleagues Will Bainbridge and Jenya Collings for their support and, in Jenya's case, her assistance in publishing the book.

We hope you find the book useful.

Chris Greenshields and Henry Weller
*CFD Direct Ltd / The OpenFOAM Foundation Ltd*

April 2022.

# Symbols

## General variables

| | |
|---|---|
| $q$, $s$ | scalars |
| $\mathbf{a}$, $\mathbf{b}$, $\mathbf{c}$ | vectors |
| $\mathbf{S}$, $\mathbf{T}$, $\mathbf{Y}$ | tensors |
| $x$, $y$, $z$ | Cartesian axis directions |
| $\mathbf{I}$ | identity tensor |
| $\boldsymbol{\Psi}$ | general parameter: scalar, vector or tensor |

## Geometry and mesh

| | |
|---|---|
| $\mathbf{C}$ | cell centre |
| $\mathbf{C}_\mathrm{f}$ | face centre |
| $\mathbf{n}$ | normal vector of unit length |
| $\mathbf{p}$ | point location |
| $S$ | surface |
| $V$ | volume |

## Numerical methods

| | |
|---|---|
| $[\,\mathbf{A}\,]$ | matrix of a matrix equation $[\,\mathbf{A}\,][\,\boldsymbol{\Psi}\,] = [\mathbf{b}]$ |
| $\mathbf{A}$ | matrix of a matrix equation $\mathbf{A} \bullet \boldsymbol{\Psi} = \mathbf{b}$ |
| $a_{i,j}$ | matrix coefficients for row $i$ and column $j$ |
| $[\mathbf{b}]$ | source vector of a matrix equation |
| $\mathbf{b}$ | source vector of a matrix equation, see above |
| $c$ | scalar coefficent |

| | |
|---|---|
| $\mathbf{D}$ | internal transformation tensor |
| $C_\Delta$ | delta coefficient $= 1/\|\mathbf{\Delta d}\|$ |
| $\mathbf{d}_\mathrm{P}$ | distance from owner cell centre to face centre |
| $\mathbf{d}_\mathrm{N}$ | distance from face centre to neighbour cell centre |
| $\mathbf{d}_\mathrm{s}$ | skewness vector |
| $f$ | numerical function |
| $\mathbf{M}$ | preconditioning matrix |
| $\mathbf{n}_\mathrm{f}$ | unit normal vector at face |
| $\mathbf{p}$ | conjugate gradient search direction |
| $r$ | ratio of consecutive gradients |
| | single-valued residual |
| $r_\mathrm{abs}$ | absolute tolerance |
| $r_\mathrm{rel}$ | relative tolerance |
| $\mathbf{r}$ | residual vector |
| $\mathbf{S}_\mathrm{f}$ | face area vector |
| $\mathbf{T}$ | transformation tensor |
| $\mathbf{R}$ | rotational tensor |
| $v$ | value fraction |
| $w$ | face weights |
| $\mathbf{Y}$ | value fraction tensor |
| $\alpha$ | under-relaxation factor |
| | conjugate gradient search magnitude |
| | GAMG correction |
| $\beta$ | advection limiter |
| $\Gamma$ | diffusivity |
| $\mathbf{\Lambda}$ | internal coefficient factor |
| $\mathbf{\Delta d}$ | centre-to-centre vector |
| $\Delta t$ | time step interval |
| $\epsilon$ | solution error |
| $\phi_\mathrm{f}$ | volumetric or mass flux at faces |
| $\phi^-$ | negative (incoming) fluxes at faces |
| $\phi^+$ | positive (outgoing) fluxes at faces |
| $\lambda$ | Crank-Nicolson offset coefficient |

**Subscripts**

| | |
|---|---|
| b | boundary value/gradient |
| D | downwind interpolation |
| ex | exact solution |
| f | values at faces |
| H | high bound value |
| L | linear interpolation |
| | low bound value |
| N | neighbour cell |
| P | owner, or current, cell |
| U | upwind interpolation |

**Superscripts**

| | |
|---|---|
| c | current stored value |
| n | new value |
| o | old time level |
| oo | old-old time level |

**Dimensionless numbers**

| | |
|---|---|
| Co | Courant number $= (\Delta t/V)\sum_f \phi^+$ |
| Pe | Péclet number $= \phi_\mathrm{f}/(\lvert\mathbf{S}_\mathrm{f}\rvert\, \alpha_\mathrm{f}\, C_\Delta)$ |

## Physical quantities (with SI units)

| | |
|---|---|
| $\mathbf{b}$ | body force per unit mass ($\mathrm{Nkg^{-1}}$) |
| $c$ | speed of sound ($\mathrm{ms^{-1}}$) |
| $c_V$ | specific heat capacity at constant volume ($\mathrm{Jkg^{-1}K^{-1}}$) |
| $D$ | mass diffusivity ($\mathrm{m^2s^{-1}}$) |
| $\mathbf{D}$ | rate of deformation tensor ($\mathrm{s^{-1}}$) |
| $e$ | specific internal energy ($\mathrm{Jkg^{-1}}$) |
| $\mathbf{f}$ | force vector (N) |
| $\mathbf{g}$ | gravitational acceleration ($\mathrm{ms^{-2}}$) |
| $h$ | heat transfer coefficient ($\mathrm{Wm^{-2}K^{-1}}$) |

| | |
|---|---|
| $\Delta h_{\mathrm{f}}$ | heat of formation per unit mass ($\mathrm{Jkg^{-1}}$) |
| $k$ | turbulent kinetic energy per unit mass ($\mathrm{Jkg^{-1}}$) |
| | wave number ($\mathrm{m^{-1}}$) |
| $K$ | specific kinetic energy ($\mathrm{Jkg^{-1}}$) |
| $L$ | characteristic length (m) |
| $l_{\mathrm{m}}$ | turbulent mixing length |
| $M$ | molar mass ($\mathrm{kgmol^{-1}}$) |
| $\mathbf{q}$ | heat per unit surface area ($\mathrm{Wm^{-2}}$) |
| $p$ | pressure (Pa) |
| | kinematic pressure ($\mathrm{m^2s^{-2}}$) |
| $p_0$ | characteristic pressure |
| $r$ | heat source per unit mass ($\mathrm{Wkg^{-1}}$) |
| | thermal source ($\mathrm{Ks^{-1}}$) |
| $R$ | specific gas constant ($\mathrm{Jkg^{-1}K^{-1}}$) |
| $\mathbb{R}$ | Universal Gas Constant ($\mathrm{JK^{-1}mol^{-1}}$) |
| $T$ | temperature (K) |
| $T^{\ominus}$ | standard temperature = 298.16K |
| $t$ | time (s) |
| $\mathcal{T}$ | characteristic time |
| $\mathbf{t}$ | traction force per unit area (Pa) |
| $\mathbf{u}$ | velocity vector ($\mathrm{ms^{-1}}$) |
| $u_x$ | velocity in $x$-direction |
| $u_\infty$ | freestream or far-field velocity |
| $U$ | characteristic speed |
| $u_{\mathrm{m}}$ | characteristic speed |
| $v$ | Kolmogorov velocity scale |
| $\mathbf{u}'$ | turbulent velocity fluctuations |
| $\mathbf{v}$ | particle velocity |
| $\overline{v}$ | mean particle velocity |
| $\mathbf{W}$ | spin tensor ($\mathrm{s^{-1}}$) |
| $\mathbf{x}$ | particle position (m) |
| $\alpha$ | thermal diffusivity $= \kappa/\rho c_v$ ($\mathrm{m^2s^{-1}}$) |
| $\alpha_e$ | energy diffusivity $= \kappa/c_v$ (Pa s) |

| | |
|---|---|
| $\gamma$ | ratio of specific heat capacities $= (R + c_V)/c_V$ |
| $\delta$ | boundary layer thickness (m) |
| $\eta$ | Kolmogorov length scale (m) |
| $\varepsilon$ | turbulent dissipation rate ($m^2s^{-3}$) |
| $\kappa$ | thermal conductivity ($Wm^{-1}K^{-1}$) |
| | Kármán's constant $\approx 0.4$ |
| $\kappa_t$ | turbulent thermal conductivity |
| $\kappa_{eff}$ | effective thermal conductivity $= \kappa + \kappa_t$ |
| $\lambda$ | mean free path (m) |
| $\mu$ | dynamic viscosity (Pa s) |
| $\nu$ | kinematic viscosity ($m^2s^{-1}$) |
| $\nu_t$ | (kinematic) turbulent viscosity |
| $\nu_{eff}$ | (kinematic) effective viscosity $= \nu + \nu_t$ |
| $\rho$ | (mass) density ($kgm^{-3}$) |
| $\boldsymbol{\sigma}$ | stress tensor (Pa) |
| $\boldsymbol{\sigma}^t$ | Reynolds stress |
| $\boldsymbol{\tau}$ | viscous stress tensor |
| $\boldsymbol{\tau}^t$ | turbulent viscous stress tensor |
| $\omega$ | specific turbulent dissipation rate ($s^{-1}$) |
| $\boldsymbol{\omega}$ | vorticity ($s^{-1}$) |

### Dimensionless numbers

| | |
|---|---|
| Eu | Euler number $= p_0/U^2$ |
| Fr | Froude number $= U/\sqrt{gL}$ |
| Kn | Knudsen number $= \lambda/L$ |
| Ma | Mach number $= \lvert\mathbf{u}\rvert/c$ |
| Pe | Péclet number $= UL/\alpha$ |
| Pr | Prandtl number $= \nu/\alpha$ |
| Re | Reynolds number $= UL/\nu$ |
| $Re_T$ | Turbulent Reynolds number $= k^2/\varepsilon\nu$ |
| Sc | Schmidt number $= \nu/D$ |
| Sr | Strouhal number $= L/(\mathcal{T}U)$ |

## Relational symbols

| | |
|---|---|
| $a = b$ | $a$ and $b$ are equal |
| $a \approx b$ | $a$ and $b$ are approximately equal |
| $a \simeq b$ | $a$ and $b$ are almost exactly equal |
| $a \equiv b$ | $a$ and $b$ are equivalent |
| $a := b$ | $a$ is assigned the value of $b$ |
| $a \propto b$ | $a$ is proportional to $b$ |
| $a \sim b$ | $a$ is of the order of magnitude of $b$ |

# Chapter 1

# Introduction

*"If we get through for two minutes only, it will be a start."*

The Jam, *Start!* (1980).

Fluid dynamics is concerned with the motion of fluids (liquids and gases) and the forces on them. "Computational" refers to computation of the flow and forces using numerical analysis. A literal definition of *computational fluid dynamics* might therefore be *the prediction of fluid motion and forces by computation using numerical analysis.*

Today, the term computational fluid dynamics, usually abbreviated to CFD, is used to describe a broader range of calculations for a wide variety of scientific and engineering applications. In particular, it is commonly used for applications involving heat, including the following:

- engines, *e.g.* internal combustion engines, turbines;
- heat recovery, *e.g.* heat exchangers;
- thermal management, *e.g.* cooling, exhaust systems;
- thermal comfort, with heating, ventilation, and air conditioning.

Thermodynamics is an important consideration in many of these applications. It relates internal energy to temperature, which affects the flow of heat. Further sources of heat include thermal radiation and chemical reactions, in particular combustion. Heat transfer may involve conduction in solid materials, coupled with the fluid flow, known as conjugate heat transfer.

If we include the examples mentioned above, a modern definition of computational fluid dynamics would be **the prediction of fluid motion and forces by computation using numerical analysis, generally extended to include heat, thermodynamics, chemistry and solids**.

Numerical analysis provides many methods and algorithms that are suitable for CFD. The methods include finite volume, finite element and finite difference, which calculate the distributions of properties, *e.g.* pressure, velocity and temperature, over regions of space which are usually fixed. Alternative methods attribute properties to particles represented by points in space, whose motions are calculated.

Particle methods are often used to approximate small scale flow features such as liquid sprays, e.g. for cooling, coating, cleaning, agriculture, food production, fire suppression, emission reduction and fuel injection. Solid particles can also be simulated in applications such as filtration, erosion, and fluidised beds.

As a presentation of general principles, this book describes numerical methodsto solve problems in fluid dynamics, up to and including heat and some basic thermodynamics, without extending into thermal radiation, chemistry and solids.

The book is solely dedicated to the *finite volume method*, the chosen method for many decades in the most popular, general-purpose CFD codes, including OpenFOAM. It also does not describe particle methods.

## 1.1 Solution overview

Let us imagine calculating fluid flow along a pipe with CFD. To perform the calculation first requires a description of the problem by:

- the domain occupied by the fluid, *i.e.* the internal region of the pipe;
- equations that represent the fluid behaviour, in terms of properties such as pressure $p$ and velocity $\mathbf{u}$;
- conditions at the boundary of the fluid domain and *initially* within the domain for the fluid properties.

This description is represented in CFD by the following:

- a computational mesh for the fluid;
- "discrete" equations and algorithms to calculate $p$ and $\mathbf{u}$;
- boundary and initial conditions for $p$ and $\mathbf{u}$.

Chapter 2 introduces the governing equations and basic models for fluid motion, forces and heat. Turbulence, which commonly occurs in many flows, is introduced in Chapter 6 and its standard modelling is described in Chapter 7.

The finite volume method is presented in Chapter 3 to express equations in discrete form using its geometric representation of a computational mesh. The algorithms used to solve the matrix equations and to couple sets of equations are given in Chapter 5.

Chapter 4 describes boundary conditions first from a numerical perspective, *i.e.* how they modify the matrix equations to influence the solution. It then covers a range of conditions that represent the behaviour at boundaries in many problems.

Internal region of pipe

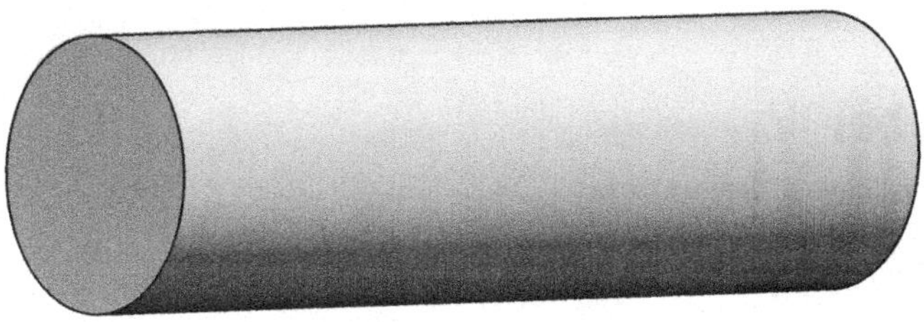

Equations (example):

$$\nabla \bullet (\mathbf{uu}) - \nabla \bullet \boldsymbol{\sigma} = -\nabla p \, , \qquad \nabla \bullet \mathbf{u} = 0$$

Computational mesh

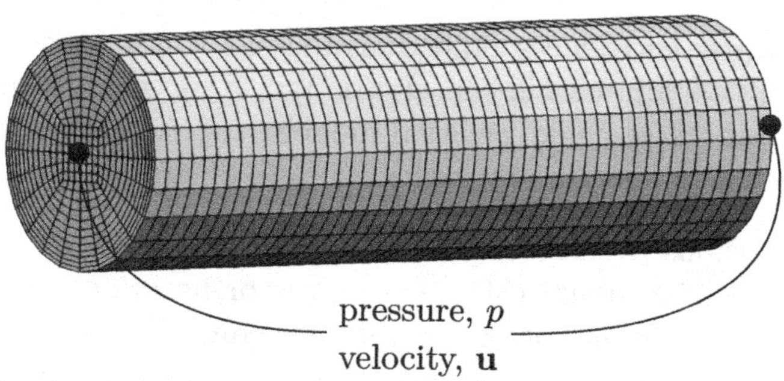


Matrix equations:

$$\begin{bmatrix} \star & \star & & \star \\ \star & \star & \star & \\ & \star & \star & \star \\ \star & & \star & \star \end{bmatrix} \begin{bmatrix} u_x \end{bmatrix} = \begin{bmatrix} \star \\ \star \\ \star \\ \star \end{bmatrix} \qquad \begin{bmatrix} \star & \star & & \star \\ \star & \star & \star & \\ & \star & \star & \star \\ \star & & \star & \star \end{bmatrix} \begin{bmatrix} u_y \end{bmatrix} = \begin{bmatrix} \star \\ \star \\ \star \\ \star \end{bmatrix}$$

$$\begin{bmatrix} \star & \star & & \star \\ \star & \star & \star & \\ & \star & \star & \star \\ \star & & \star & \star \end{bmatrix} \begin{bmatrix} u_z \end{bmatrix} = \begin{bmatrix} \star \\ \star \\ \star \\ \star \end{bmatrix} \qquad \begin{bmatrix} \star & \star & & \star \\ \star & \star & \star & \\ & \star & \star & \star \\ \star & & \star & \star \end{bmatrix} \begin{bmatrix} p \end{bmatrix} = \begin{bmatrix} \star \\ \star \\ \star \\ \star \end{bmatrix}$$

# Chapter 2

# Fluid Dynamics

*"Stands on shifting sands, the scales held in her hands. The wind it just whips her and wails and fills up her brigantine sails."*

The Stone Roses, *Waterfall* (1989).

This chapter introduces fluid dynamics for CFD. It describes: governing equations, *i.e.* conservation of mass, momentum and energy; and, associated physical models, *e.g.* for viscosity, heat conduction and thermodynamics.

The equations describe fluid motion, forces and heat in time and three-dimensional (3D) space. Vector notation provides a mathematical framework to present the equations in a compact form. It enables the equations to be presented independently of any co-ordinate system, *e.g.* Cartesian ($x/y/z$) or spherical ($r/\theta/\varphi$). It includes a standard set of algebraic operations, *e.g.* the inner (dot) and outer products.

The notation helps to ensure that the terms in equations are unchanged, or *invariant*, under a co-ordinate system transformation. Without invariance, a flow solution, *e.g.* along a pipe, would be dependent on the orientation of the pipe with respect to the co-ordinate system. Logically this dependence cannot exist; the laws of motion are the same in all "inertial frames".[†]

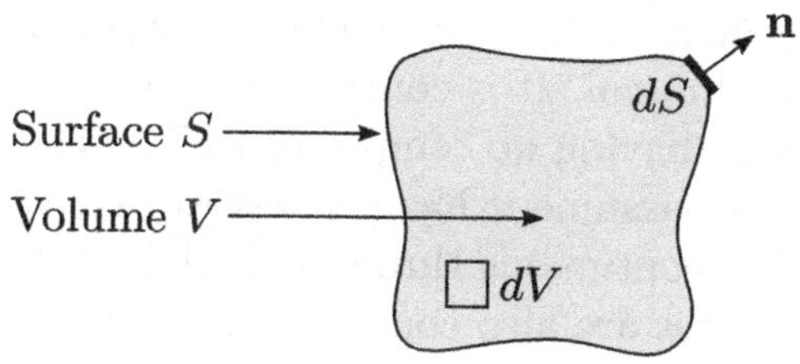


The derivation of the governing equations uses a control volume $V$ bounded by a surface $S$, presented using the two-dimensional (2D) illustration above. We use $dV$ and $dS$ to describe an infinitesimally small volume and surface, respectively, and $\mathbf{n}$ is the unit normal vector for each increment of surface $dS$, discussed in Sec. 2.1. It is important to note in any derivation whether the volume is defined as fixed in space or moving with the fluid.

Each derivation generally begins with a definite integral of some quantity, *e.g.* $\boldsymbol{\Psi}$, over the volume $V$ denoted by

$$\int_V \boldsymbol{\Psi}\, dV. \tag{2.1}$$

If this notation is unfamiliar, understand it to mean a summation for all increments of volume $dV$ that make up the total volume $V$. The summed values are $\boldsymbol{\Psi} \times dV$, where $\boldsymbol{\Psi}$ is the value in the respective $dV$.

The derivations also use integrals over the surface $S$, *e.g.*

$$\int_S \mathbf{n}\boldsymbol{\Psi}\, dS \quad \text{or} \quad \int_S (d\mathbf{S}\, \boldsymbol{\Psi}), \tag{2.2}$$

where $d\mathbf{S} = \mathbf{n}\, dS$. Volume and surface integrals are connected through Gauss's Theorem, introduced in Sec. 2.4.

[†]Galileo Galilei, *Dialogo sopra i due massimi sistemi del mondo*, 1632.

## 2.1 Pressure

The equations of fluid dynamics in CFD treat the fluid as a continuous medium, or *continuum*. It is continuous in the sense that we consider the fluid as having no "empty space" by ignoring its molecular nature. We assume it has properties that vary from point to point and are continuous throughout the solution domain, and whose derivatives are also continuous.

Pressure is an important property of a fluid, denoted by $p$. It describes the amount of force per unit surface area which *acts on* a surface, in the direction *perpendicular* to the surface.

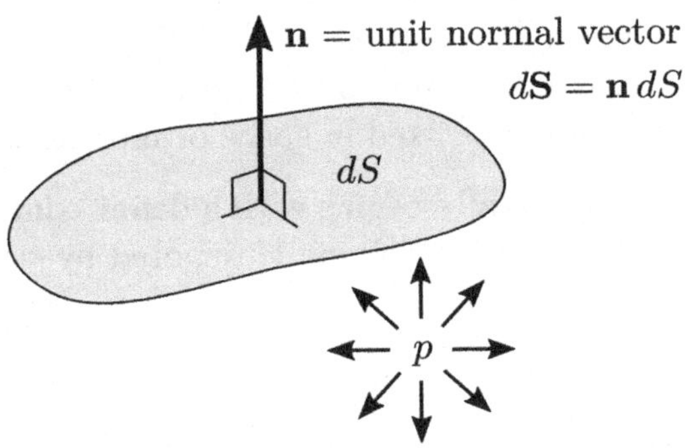


Pressure $p$ is a *scalar* that produces a force *vector* with direction normal to the surface. When pressure is applied to one side of a segment of surface with area $dS$, the force $d\mathbf{f}$ points *away* from that side by

$$d\mathbf{f} = (\mathbf{n}\, dS)\, p = d\mathbf{S}\, p, \tag{2.3}$$

where $\mathbf{n}$ is the vector of *unit length*, normal to the surface $dS$. The term *vector* denotes a geometric entity with magnitude and direction; a surface can be represented by a *surface area vector* $d\mathbf{S} = \mathbf{n}\, dS$ of magnitude $dS$ and direction $\mathbf{n}$.

While pressure exerts a force *on a surface*, the fluid experiences a force which is compressive in nature, assuming $p$ is positive.

Pressure is measured in SI units of pascal Pa $\equiv$ kgm$^{-1}$s$^{-2}$. From Sec. 2.13 onwards, however, we generally use $p$ to represent the *kinematic* pressure, in units m$^2$s$^{-2}$, obtained by dividing the true pressure by a constant density $\rho$.

## Scalar fields

The majority of properties, *e.g.* pressure, temperature, energy, density, volume, *etc.*, can be represented by a single number, or *scalar*. A *scalar field* describes a scalar property, *e.g.* pressure, which varies from point to point across some spatial domain.

Point locations can be defined in any co-ordinate system of axes, *e.g.* Cartesian ($x$, $y$, $z$), and in any orientation. A scalar field is *invariant*, meaning the scalar values are the same irrespective of the co-ordinate system used.

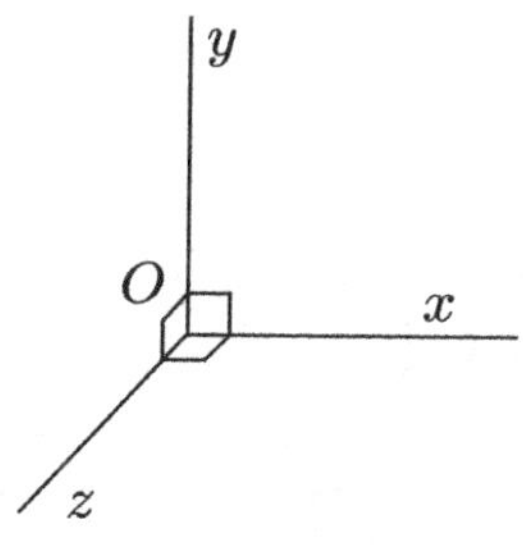


In this book, space and fields will be described in a co-ordinate system with right-handed rectangular Cartesian axes. The axes are constructed by defining an origin $O$ from which three lines are drawn at right angles to each other, termed the $Ox$, $Oy$, $Oz$ axes. A right-handed set of axes requires that looking down the $Oz$ axis with $O$ nearest, an arc from the $Ox$ axis to the $Oy$ axis is in a clockwise sense.

## 2.2 Velocity

Like force, velocity $\mathbf{u}$ is a vector with direction and magnitude, with SI units of $\mathrm{ms}^{-1}$. Using the vector $\mathbf{x}(t)$ to denote the position of a particle of fluid, its velocity is

$$\mathbf{u} = \frac{d\mathbf{x}}{dt} = \lim_{\Delta t \to 0} \frac{\Delta \mathbf{x}}{\Delta t}. \tag{2.4}$$

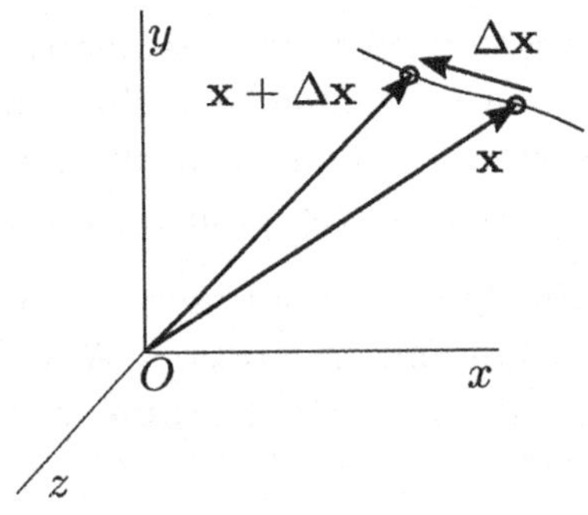


### Vector fields

While $\mathbf{u}$ can be used to denote a single velocity with magnitude and direction, it can also denote a *vector field* of velocity which varies from point to point across a spatial domain. A vector is represented by 3 numbers, relating to the co-ordinate system being used, *e.g.* $x$, $y$, $z$, in the Cartesian system.

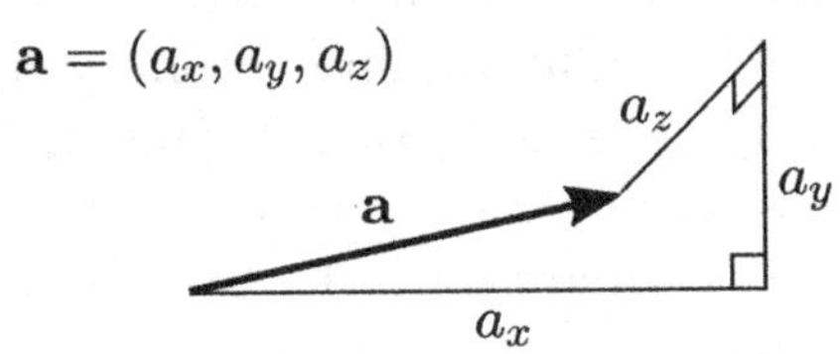

While the magnitude and direction of a vector is fixed, it is not invariant since the 3 *values* depend on the co-ordinate system used. We represent a vector without reference to the co-ordinate system by bold text, *e.g.* "**a**" (compared to scalar "$a$").

## Basic vector algebra

Addition and subtraction of 2 vectors is performed by operating on respective components. Subtraction of two vectors **a** and **b** is performed by

$$\mathbf{a} - \mathbf{b} = (a_x - b_x, a_y - b_y, a_z - b_z) .$$

Multiplication of any vector **a** by a scalar $s$ is performed by multiplying all the components by the scalar, *e.g.*

$$s\mathbf{a} = (sa_x, sa_y, sa_z)$$

Addition and multiplication are *commutative*, *i.e.* variables can be in any order ($s\mathbf{a} = \mathbf{a}s$). Subtraction is not commutative. Products between scalars and vectors are *distributive*, *i.e.*

$$s(\mathbf{a} + \mathbf{b}) = s\mathbf{a} + s\mathbf{b},$$

and with additional scalar $q$,

$$(s + q)\mathbf{a} = s\mathbf{a} + q\mathbf{a}.$$

Division between a vector **a** and a scalar is only relevant when the scalar is the second argument of the operation, *i.e.*

$$\mathbf{a}/s = (a_x/s, a_y/s, a_z/s) .$$

## 2.3 Flow through a surface

The concept of flow through a surface appears in many areas of CFD, including fluid dynamics equations, numerical methods, boundary conditions and general flow calculations. When we talk about something that travels through a surface, the term *flux* is generally used.†

To quantify the flux of some property, we multiply the area of surface by the property at the surface. If the property is a vector, we take the component *normal to the surface*. For example, the flux $d\phi$ associated with velocity through a surface segment of area $dS$ would be $d\phi = u_n \, dS$.

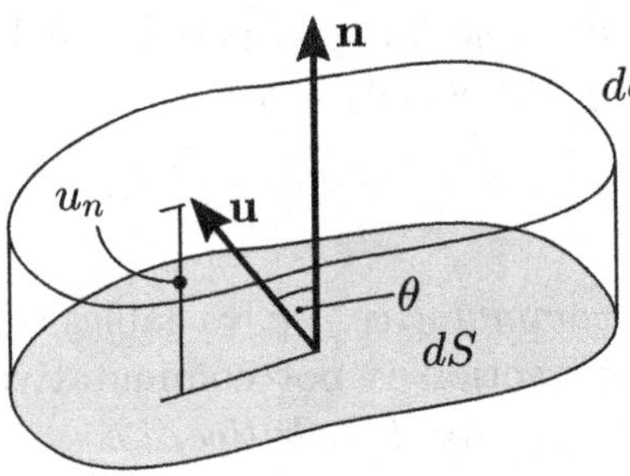


$$\begin{aligned} d\phi &= u_n \, dS = \mathbf{n} \bullet \mathbf{u} \, dS = d\mathbf{S} \bullet \mathbf{u} \\ &= \text{volume (of box) per second} \end{aligned}$$

As shown in the figure, $u_n$ can be calculated from $\mathbf{u}$ by the inner product with normal vector $\mathbf{n}$ of unit length, expressed as

$$u_n = \mathbf{n} \bullet \mathbf{u}. \tag{2.5}$$

The flux associated with $\mathbf{u}$ is

$$d\phi = \mathbf{n} \bullet \mathbf{u} \, dS = d\mathbf{S} \bullet \mathbf{u}. \tag{2.6}$$

It is a good habit to write $d\mathbf{S}$ first since the order matters with a tensor, *e.g.* stress $\boldsymbol{\sigma}$, introduced in Sec. 2.6, *i.e.* $\boldsymbol{\sigma} \bullet d\mathbf{S} \neq d\mathbf{S} \bullet \boldsymbol{\sigma}$.

† Joseph Fourier used the term *fluxion* in relation to flow of heat in *Théorie analytique de la chaleur*, 1822; James Clerk Maxwell used the term *flux* in *A Treatise on Electricity and Magnetism*, 1873.

## Inner product of two vectors

The normal component of velocity $u_n$ is described in Eq. (2.5) by the inner, or "dot", product of $\mathbf{u}$ and $\mathbf{n}$. It is calculated for vectors $\mathbf{a}$ and $\mathbf{b}$ as shown in the figure below, where $|\ldots|$ denotes the magnitude of the vector and $\theta$ is the internal angle between the two vectors.

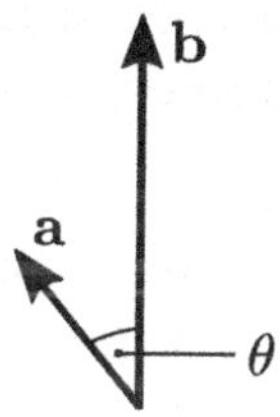


$$\mathbf{a} \bullet \mathbf{b} = |\mathbf{a}||\mathbf{b}| \cos\theta$$

The inner product of two vectors is a scalar invariant, since the magnitudes and angle are the same irrespective of the coordinate system used. It is calculated from vector components as follows:

$$\mathbf{a} \bullet \mathbf{b} = a_x b_x + a_y b_y + a_z b_z.$$

The inner product of two vectors is commutative, *i.e.* $\mathbf{a} \bullet \mathbf{b} = \mathbf{b} \bullet \mathbf{a}$. It is distributive, *i.e.* with an additional vector $\mathbf{c}$,

$$\mathbf{a} \bullet (\mathbf{b} + \mathbf{c}) = \mathbf{a} \bullet \mathbf{b} + \mathbf{a} \bullet \mathbf{c}.$$

Scalar multiplication and inner products are associative, *i.e.*

$$s(\mathbf{a} \bullet \mathbf{b}) = s\mathbf{a} \bullet \mathbf{b} = \mathbf{a} \bullet s\mathbf{b} = (\mathbf{a} \bullet \mathbf{b})s.$$

An inner product of a vector with itself is simply the square of the vector magnitude, *i.e.*

$$\mathbf{a} \bullet \mathbf{a} = |\mathbf{a}|^2. \tag{2.7}$$

## 2.4 Conservation of mass

The law of conservation of mass can be written

$$\boxed{\frac{\partial \rho}{\partial t} + \nabla \cdot (\rho \mathbf{u}) = 0} \tag{2.8}$$

where $\rho$ is the mass density of the fluid. The equation can be derived by considering a volume $V$ *fixed in space* (note !!), bounded by a surface $S$. If the volume is filled by a fluid with density $\rho$, its mass is $M = \int_V \rho \, dV$.

The rate of increase of mass inside the volume must match the rate of inflow of mass across the volume's surface. The latter is calculated by integrating the mass flux $\rho\phi$ over the surface, noting the negative sign due to $d\mathbf{S}$ pointing out of the volume.

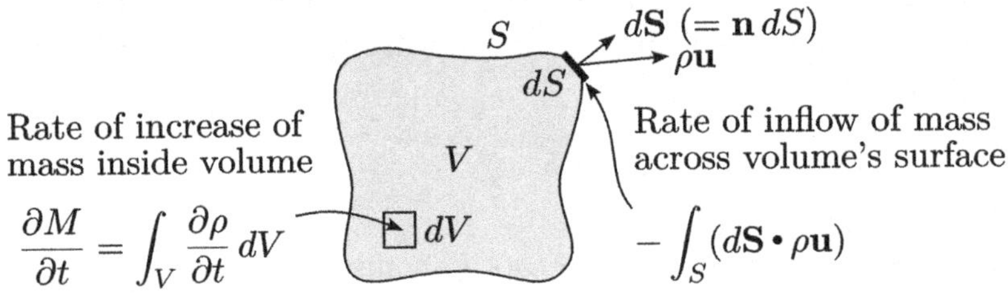


Gauss's Theorem relates surface and volume integrals by

$$\int_S (d\mathbf{S} \cdot \rho \mathbf{u}) = \int_V \nabla \cdot (\rho \mathbf{u}) \, dV. \tag{2.9}$$

Equating the rate of mass increase to rate of inflow and applying Eq. (2.9) gives

$$\int_V \left[ \frac{\partial \rho}{\partial t} + \nabla \cdot (\rho \mathbf{u}) \right] dV = 0. \tag{2.10}$$

Since the integral is valid for any volume $V$, it follows that the integrand (in [...]) must equal 0, resulting in Eq. (2.8).

## Divergence

Divergence, denoted by $\nabla \bullet (\ldots)$, indicates the tendency of a vector field to point outward of a closed surface. For example, When the divergence of velocity $\nabla \bullet \mathbf{u}$ is positive, the fluid is expanding; negative divergence indicates contraction. Imagine a volume of fluid with a complex distribution of $\mathbf{u}$ at its bounding surface, below left. The $\nabla \bullet \mathbf{u}$ calculation will isolate the diverging component (right) from uniform flow (centre).

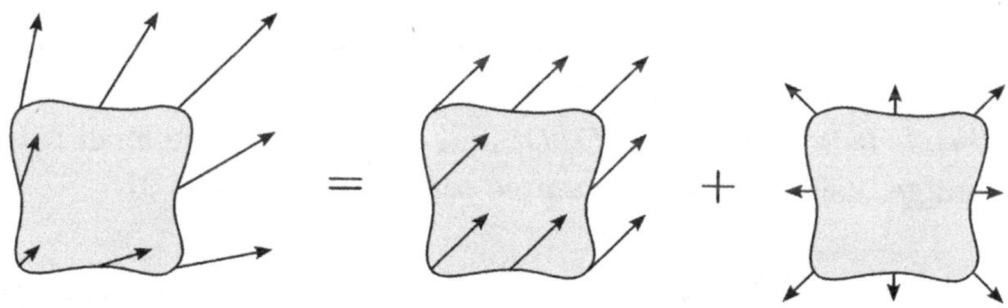

The divergence of velocity is calculated by integrating — *i.e.* summing — fluxes over the closed surface. To define divergence *at a point*, we consider the limiting case where the volume tends to zero. For a surface $S$ that encloses a volume $\Delta V$, divergence is the flux across the surface per unit volume, as $\Delta V \to 0$, *e.g.*

$$\nabla \bullet \mathbf{u} = \lim_{\Delta V \to 0} \frac{1}{\Delta V} \int_S (d\mathbf{S} \bullet \mathbf{u}). \tag{2.11}$$

## The nabla operator

The nabla symbol $\nabla$ can be considered a *vector operator*

$$\nabla \equiv \left( \frac{\partial}{\partial x}, \frac{\partial}{\partial y}, \frac{\partial}{\partial z} \right). \tag{2.12}$$

Divergence is the inner product with $\nabla$, *e.g.*

$$\nabla \bullet \mathbf{u} = \frac{\partial u_x}{\partial x} + \frac{\partial u_y}{\partial y} + \frac{\partial u_z}{\partial z}. \tag{2.13}$$

## 2.5 Time derivatives

The conservation of mass Eq. (2.8) included a partial derivative in time $\partial\rho/\partial t$ relating to a *fixed region of space.* This is the *local rate of change of* $\rho$, relating to the change in $\rho$ in the fluid measured by an observer at a fixed location. It is not the time rate of change experienced by a mass of fluid particles as they move through space. In the same way, $\partial\mathbf{u}/\partial t$ is not the acceleration experienced by the fluid.

Acceleration relates to the *material*, or *substantive*, derivative which describes the time rate of change of a *fixed mass of moving material.* It is denoted by $D/Dt$ and is related to the local rate of change, using $\boldsymbol{\Psi}$ as an example tensor of any rank, by

$$\boxed{\frac{D\boldsymbol{\Psi}}{Dt} = \frac{\partial\boldsymbol{\Psi}}{\partial t} + \mathbf{u}\cdot\nabla\boldsymbol{\Psi}} \tag{2.14}$$

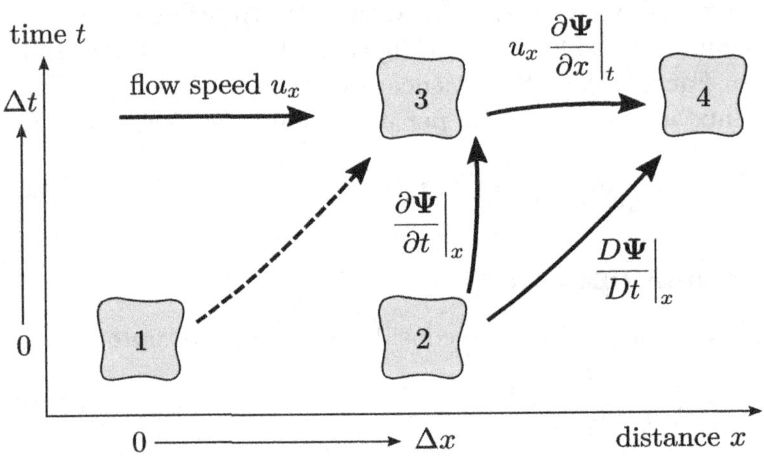


The relation is derived from the *chain rule of differential.* In one dimension, it is illustrated by two particles of fluid that occupy positions 1 and 2 at some initial time, then positions 3 and 4

at a later time $\Delta t$. The particles move in the $x$-direction at speed $u_x = dx/dt$, such that the particle at 1 later occupies the position 3 and the particle at 2 occupies position 4.

The material time derivative of $\mathbf{\Psi}$, following the mass from 2→4, is the sum of: the local change in $\mathbf{\Psi}$ at a fixed position $x$ (2→3); and, the change due to the *gradient* of $\mathbf{\Psi}$ between positions 3 and 4, fixing time $t$. This equates to

$$\left.\frac{D\mathbf{\Psi}}{Dt}\right|_x = \left.\frac{\partial \mathbf{\Psi}}{\partial t}\right|_x + \frac{dx}{dt}^{=u_x} \left.\frac{\partial \mathbf{\Psi}}{\partial x}\right|_t . \tag{2.15}$$

The material derivative relation in Eq. (2.14) is simply the 3D equivalent of Eq. (2.15).

## Gradient

The last term in Eq. (2.14) introduces the *gradient* denoted by $\nabla\mathbf{\Psi}$. If $\mathbf{\Psi}$ is a scalar, the gradient produces a vector whose magnitude and direction is that of the *steepest gradient.*

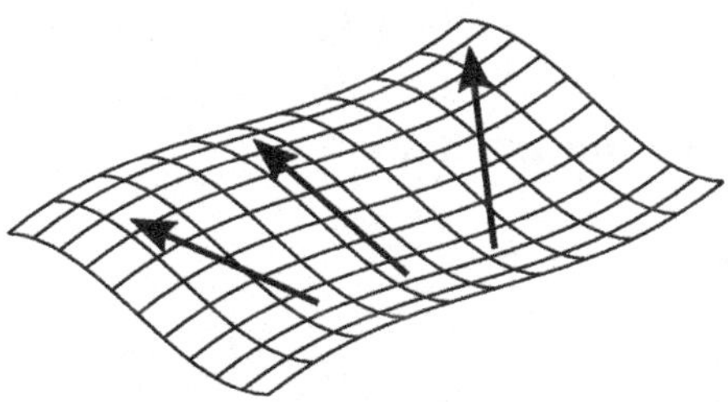

The figure above illustrates the gradient using a surface that represents a distribution of a scalar field in 2 directions. The gradients at 3 locations are in the direction of steepest ascent.

When $\mathbf{\Psi}$ is a vector, the gradient produces a tensor, representing the direction and magnitude of steepest ascent for each of the 3 components of the vector.

## 2.6 Forces at a surface

The next law to define is that of conservation of momentum, *i.e.* Newton's second law of motion ($\mathbf{f} = m\mathbf{a}$) for fluids. It involves forces within the fluid so requires a description of forces at a surface $S$ bounding a volume $V$.

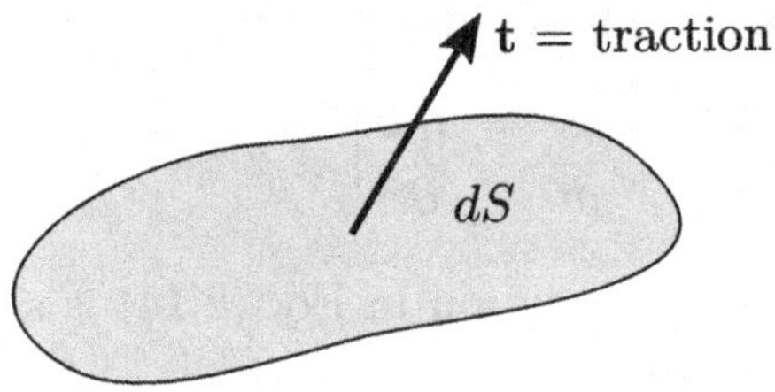


The force $d\mathbf{f}$ is in the direction of the traction vector with magnitude of (traction × surface area) — compare with Eq. (2.3)

$$d\mathbf{f} = \mathbf{t}\, dS \tag{2.16}$$

The tractions at a surface bounding some volume of fluid (or any continuum, *e.g.* solid) *depend on the orientation of the surface.* Consequently, we cannot define forces within a fluid as traction vectors *at points* within the fluid.

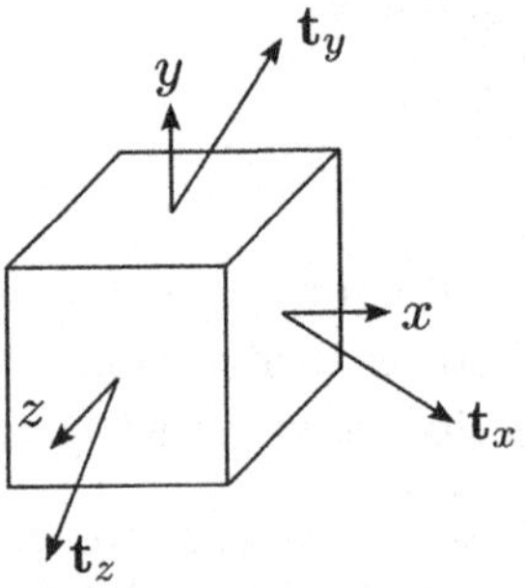


$$\boldsymbol{\sigma} = \begin{pmatrix} \mathbf{t}_x \\ \mathbf{t}_y \\ \mathbf{t}_z \end{pmatrix} = \begin{pmatrix} \sigma_{xx} & \sigma_{xy} & \sigma_{xz} \\ \sigma_{yx} & \sigma_{yy} & \sigma_{yz} \\ \sigma_{zx} & \sigma_{zy} & \sigma_{zz} \end{pmatrix}$$

Instead, we require 3 *traction vectors* $\mathbf{t}_x$, $\mathbf{t}_y$ and $\mathbf{t}_z$, defined in planes perpendicular to one another, *i.e.* $x$, $y$ and $z$. This results in the *stress tensor* $\boldsymbol{\sigma}$ with 9 components, consisting of 3 traction vectors, each containing 3 components.

The traction can be calculated at a surface with any orientation by taking the inner product of the unit normal vector $\mathbf{n}$ and the stress tensor $\boldsymbol{\sigma}$ such that

$$d\mathbf{f} = \mathbf{t}\, dS = (\mathbf{n} \bullet \boldsymbol{\sigma})\, dS = d\mathbf{S} \bullet \boldsymbol{\sigma}. \tag{2.17}$$

## Tensors

We have defined the stress tensor†, an entity with 9 component values, corresponding to our $x$, $y$ and $z$ axes (or more specifically *base vectors of unit length* aligned with our $x$, $y$ and $z$ axes).

In fact, the term "tensor" describes any entity with multiple component values corresponding to the dimensions of space — here 3. A tensor has *rank* $r$, such that the number of component values for 3D space $= 3^r$.

In this book we use the term "tensor" to mean "tensor of rank 2" unless otherwise noted. A vector is a tensor of rank 1 and a scalar is rank 0.

The inner product of a vector $\mathbf{a}$ and tensor $\mathbf{T}$ produces a vector whose 3 components are evaluated as follows:

$$\mathbf{a} \bullet \mathbf{T} = \begin{pmatrix} a_x T_{xx} + a_y T_{yx} + a_z T_{zx} \\ a_x T_{xy} + a_y T_{yy} + a_z T_{zy} \\ a_x T_{xz} + a_y T_{yz} + a_z T_{zz} \end{pmatrix} \tag{2.18}$$

This inner product tensor is commutative only if $\mathbf{T}$ is symmetric since $\mathbf{a} \bullet \mathbf{T} = \mathbf{T}^{\mathrm{T}} \bullet \mathbf{a}$ — see Eq. (2.24) for the *transpose* $\mathbf{T}^{\mathrm{T}}$.

---

†More precisely the "Cauchy stress tensor", introduced by Augustin-Louis Cauchy in *De la pression ou tension dans un corps solide*, 1827.

## 2.7 Conservation of momentum

The law of conservation of momentum can be written[†]

$$\boxed{\rho \frac{D\mathbf{u}}{Dt} = \nabla \cdot \boldsymbol{\sigma} + \rho \mathbf{b}} \tag{2.19}$$

where $\mathbf{b}$ represents any body force per unit mass. Body forces represent any force which does not act at a bounding surface, including those that act at a distance, such as gravitational force.

The equation is derived by considering the time rate of change of the momentum of a mass of particles. We consider a *volume of material of fixed mass* moving through space and therefore present rate of change by the material derivative $D/Dt$.

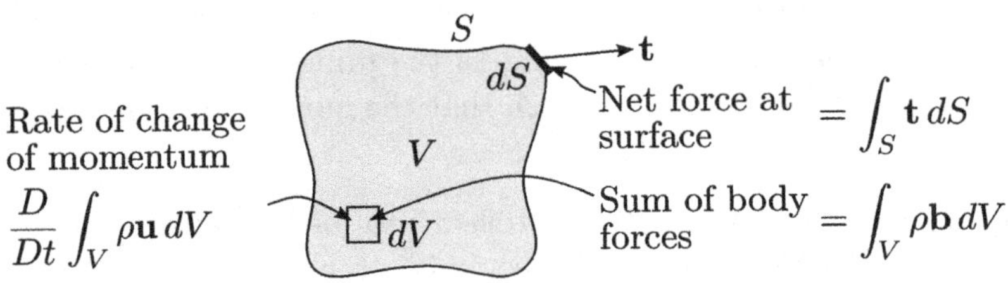


Applying Gauss's theorem and Eq. (2.16), the surface force is

$$\int_S \mathbf{t}\, dS = \int_S \mathbf{n} \cdot \boldsymbol{\sigma}\, dS = \int_S (d\mathbf{S} \cdot \boldsymbol{\sigma}) = \int_V \nabla \cdot \boldsymbol{\sigma}\, dV\,. \tag{2.20}$$

Equating the rate of change of momentum to the forces and, noting mass is fixed so $\rho\, dV$ is constant in time, gives

$$\int_V \left[ \rho \frac{D\mathbf{u}}{Dt} - \nabla \cdot \boldsymbol{\sigma} - \rho \mathbf{b} \right] dV = 0\,. \tag{2.21}$$

The integrand must equal 0, resulting in Eq. (2.19).

[†]Cauchy's first law of motion (1827)

Divergence was described in section 2.4 as the flux across a surface per unit volume as $\Delta V \to 0$. The divergence of stress similarly represents the stress flux across the surface, *i.e.* force, per unit volume as $\Delta V \to 0$, given by

$$\nabla \cdot \boldsymbol{\sigma} = \lim_{\Delta V \to 0} \frac{1}{\Delta V} \int_S (d\mathbf{S} \cdot \boldsymbol{\sigma}). \tag{2.22}$$

Eq. (2.19) specifically relates to *linear* momentum. Instead, conservation of *angular* momentum†, in the absence of any "couple stresses" that generate a moment field, is given by

$$\boldsymbol{\sigma} = \boldsymbol{\sigma}^{\mathrm{T}} \quad (\boldsymbol{\sigma} \text{ is symmetric}). \tag{2.23}$$

The derivation of Eq. (2.23) is fairly complex so is omitted here.

## Tensor symmetry

Symmetry in a tensor $\mathbf{T}$ refers to components being symmetric about the diagonal, *i.e.* $T_{12} = T_{21}$, $T_{13} = T_{31}$ and $T_{23} = T_{32}$. The *transpose* of a tensor, denoted by the 'T' superscript, switches components across the diagonal such that:

$$\mathbf{T}^{\mathrm{T}} = \begin{pmatrix} T_{xx} & T_{yx} & T_{zx} \\ T_{xy} & T_{yy} & T_{zy} \\ T_{xz} & T_{yz} & T_{zz} \end{pmatrix} \tag{2.24}$$

$\mathbf{T}$ is therefore *symmetric* if $\mathbf{T} = \mathbf{T}^{\mathrm{T}}$. A *skew* (anti-symmetric) tensor has $\mathbf{T} = -\mathbf{T}^{\mathrm{T}}$. A tensor can be decomposed into symmetric and skew parts by:

$$\mathbf{T} = \underbrace{\frac{1}{2}(\mathbf{T} + \mathbf{T}^{\mathrm{T}})}_{\textit{symmetric}} + \underbrace{\frac{1}{2}(\mathbf{T} - \mathbf{T}^{\mathrm{T}})}_{\textit{skew}} = \operatorname{sym}\mathbf{T} + \operatorname{skw}\mathbf{T} \tag{2.25}$$

† Cauchy's second law of motion (1827)

## 2.8 Flow in a volume

We presented conservation of momentum, Eq. (2.19), in terms of the material derivative $D\mathbf{u}/Dt$. To solve the equation in this form, would require a method that tracks particles of fluid as they move around.

Generally it is easier to solve the equation by *fixing* the volume of space and solving for the fluid motion through it. To enable this, we replace the material derivative for $\mathbf{u}$ using the following expression:

$$\boxed{\rho\frac{D\mathbf{u}}{Dt} = \frac{\partial \rho\mathbf{u}}{\partial t} + \nabla \cdot (\rho\mathbf{u}\mathbf{u})} \tag{2.26}$$

Eq. (2.26) is derived from the material derivative Eq. (2.14) by

$$\rho\frac{D\mathbf{u}}{Dt} = \underbrace{\rho\frac{\partial \mathbf{u}}{\partial t}}_{\frac{\partial \rho\mathbf{u}}{\partial t} - \mathbf{u}\cancel{\frac{\partial \rho}{\partial t}}} + \underbrace{\rho\mathbf{u}\cdot\nabla\mathbf{u}}_{\nabla\cdot(\rho\mathbf{u}\mathbf{u}) - \mathbf{u}\cancel{\nabla\cdot(\rho\mathbf{u})}}$$

$$\frac{\partial \rho\mathbf{u}}{\partial t} - \mathbf{u}\cancel{\frac{\partial \rho}{\partial t}} + \nabla\cdot(\rho\mathbf{u}\mathbf{u}) - \mathbf{u}\cancel{\nabla\cdot(\rho\mathbf{u})}$$

$$\text{(cancelled terms)} = 0 \text{ by Eq. (2.8)}$$

Conservation of momentum can then be written

$$\frac{\partial \rho\mathbf{u}}{\partial t} + \nabla \cdot (\rho\mathbf{u}\mathbf{u}) = \nabla \cdot \boldsymbol{\sigma} + \rho\mathbf{b}. \tag{2.27}$$

A term of the form $\nabla \cdot (\mathbf{u}\boldsymbol{\Psi})$ or $\nabla \cdot (\rho\mathbf{u}\boldsymbol{\Psi})$ represents the bulk motion of the fluid at velocity $\mathbf{u}$, which transports the property $\boldsymbol{\Psi}$ (a tensor of any rank) by advection†. For example, if the property $\boldsymbol{\Psi}$ is temperature, then advection will transport heat from one region of the flow domain to another.

†The term "convection" is sometimes used, but it has conflicting meanings, so we avoid it here.

Advection contains a divergence derivative so represents a flux across a surface per unit volume. The example in Eq. (2.27) represents the flux of momentum, where the advected property is itself $\mathbf{u}$ (or $\rho\mathbf{u}$, depending whether $\rho$ is associated with bulk flow or the advected property).

## Outer product of two vectors

The advection term in Eq. (2.27) includes a product of two vectors $\mathbf{uu}$. This is the outer product of two vectors[†] which produces a tensor by

$$\mathbf{ab} = \begin{pmatrix} a_x b_x & a_x b_y & a_x b_z \\ a_y b_x & a_y b_y & a_y b_z \\ a_z b_x & a_z b_y & a_z b_z \end{pmatrix}. \tag{2.28}$$

## Inner product of two tensors

The inner product of two tensors, *e.g.* $\mathbf{T} \bullet \mathbf{S}$ produces a tensor $\mathbf{Y}$ where the components are (replacing $i, j$ with $x, y, z$)

$$Y_{ij} = T_{ix} S_{xj} + T_{iy} S_{yj} + T_{iz} S_{zj}. \tag{2.29}$$

## Identity tensor

The identity tensor $\mathbf{I}$ is the tensor equivalent of unity (one) such that for any tensor $\mathbf{T} = \mathbf{I} \bullet \mathbf{T} = \mathbf{T} \bullet \mathbf{I}$,

$$\mathbf{I} = \begin{pmatrix} 1 & 0 & 0 \\ 0 & 1 & 0 \\ 0 & 0 & 1 \end{pmatrix}. \tag{2.30}$$

---

[†]sometimes written "$\mathbf{a} \otimes \mathbf{b}$", but we write it as a product with no symbol, similar to a scalar multiplication.

## 2.9 Conservation and boundedness

The material derivative Eq. (2.14) includes the $\mathbf{u} \cdot \nabla \boldsymbol{\Psi}$ term. However, the term can take for the form $\nabla \cdot (\mathbf{u}\boldsymbol{\Psi})$, by applying conservation of mass Eq. (2.8), as shown in Sec. 2.8. The two terms are related by the product rule

$$\nabla \cdot (\mathbf{u}\boldsymbol{\Psi}) = (\nabla \cdot \mathbf{u})\boldsymbol{\Psi} + \mathbf{u} \cdot \nabla \boldsymbol{\Psi}. \tag{2.31}$$

While the two terms $\nabla \cdot (\mathbf{u}\boldsymbol{\Psi})$ and $\mathbf{u} \cdot \nabla \boldsymbol{\Psi}$ appear similar, they affect the solution of an equation is different ways.

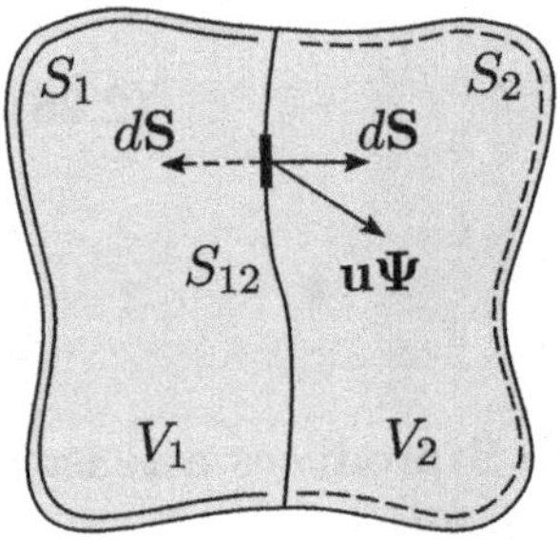


$$\begin{aligned} &\int_{V_1} \nabla \cdot (\mathbf{u}\boldsymbol{\Psi})\, dV + \int_{V_2} \nabla \cdot (\mathbf{u}\boldsymbol{\Psi})\, dV \\ &= \int_{S_1+S_{12}} (d\mathbf{S} \cdot \mathbf{u}\boldsymbol{\Psi}) + \int_{S_2+S_{12}} (d\mathbf{S} \cdot \mathbf{u}\boldsymbol{\Psi}) \\ &= \int_{S_1+S_2} (d\mathbf{S} \cdot \mathbf{u}\boldsymbol{\Psi}) = \int_{V} \nabla \cdot (\mathbf{u}\boldsymbol{\Psi})\, dV \end{aligned}$$

The $\nabla \cdot (\mathbf{u}\boldsymbol{\Psi})$ term *maintains conservation* in $\boldsymbol{\Psi}$. All variables are inside (to the right of) the divergence $\nabla \cdot (\ldots)$, so when we integrate it over a volume $V$, it can be entirely transformed by Gauss's theorem to an integral of the flux of $\mathbf{u}\boldsymbol{\Psi}$ at $S$.

If we split $V$ into two sub-volumes $V_1$ and $V_2$, the surface integrals are also split over two surfaces $S_1 + S_{12}$ and $S_2 + S_{12}$, where $S_{12}$ is the part of the surface common to both volumes.

The fluxes at $S_{12}$ for each sub-volume have equal magnitude but opposite sign, since their respective surface area vectors $d\mathbf{S}$ point outwards from the volume in opposing directions. These fluxes cancel one another out, such that the sum of integrals over $V_1$ and $V_2$ is identically equal to the integral over $V = V_1 + V_2$.

Thus, the term $\nabla \cdot (\mathbf{u}\boldsymbol{\Psi})$ ensures conservation in $\boldsymbol{\Psi}$ across a surface separating regions of the flow domain. Conservation is guaranteed at all points in the limit $\Delta V \rightarrow 0$ for any sub-volume.

The $\mathbf{u} \cdot \nabla \boldsymbol{\Psi}$ form cannot transform to a surface integral so does not ensure conservation. Instead it *ensures boundedness*, as demonstrated by the solution of $\boldsymbol{\Psi}$ in the following equation in one ($x$) dimension (1D)

$$\frac{\partial \boldsymbol{\Psi}}{\partial t} + u_x \frac{\partial \boldsymbol{\Psi}}{\partial x} = 0. \tag{2.32}$$

The solution to this equation has the form $\boldsymbol{\Psi} = f(x - u_x t)$. If $u_x$ is uniform, then the profile of $\boldsymbol{\Psi}$ does not change but simply translates at a speed $u_x$ as shown in the left diagram. If $u_x$ varies with $x$, the profile changes, *e.g.* flattens, as shown in the right diagram.

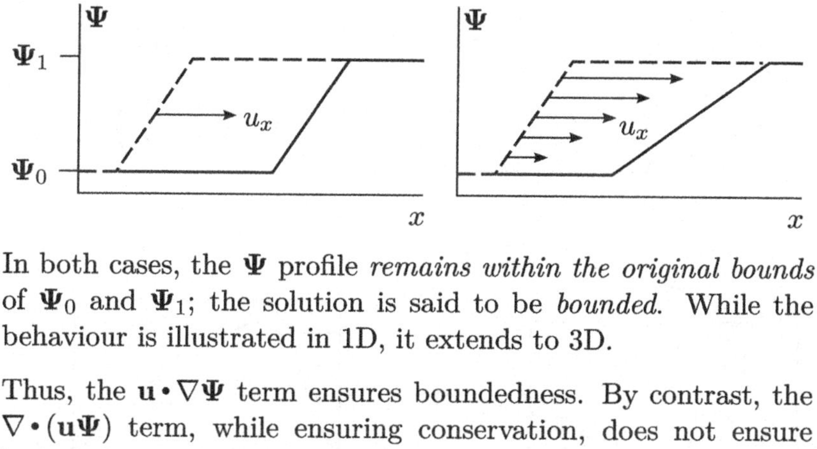


In both cases, the $\boldsymbol{\Psi}$ profile *remains within the original bounds* of $\boldsymbol{\Psi}_0$ and $\boldsymbol{\Psi}_1$; the solution is said to be *bounded*. While the behaviour is illustrated in 1D, it extends to 3D.

Thus, the $\mathbf{u} \cdot \nabla \boldsymbol{\Psi}$ term ensures boundedness. By contrast, the $\nabla \cdot (\mathbf{u}\boldsymbol{\Psi})$ term, while ensuring conservation, does not ensure boundedness when $\nabla \cdot \mathbf{u} \neq 0$. The two terms are connected by $(\nabla \cdot \mathbf{u})\boldsymbol{\Psi}$ which changes $\boldsymbol{\Psi}$ due to expansion/contraction of the fluid, as discussed in Sec. 2.4. Its effect is illustrated by flattening of the profile in the right diagram, since a non-uniform $u_x$ corresponds to $\nabla \cdot \mathbf{u} \neq 0$.

## 2.10 Fluid deformation

Conservation of momentum Eq. (2.19) includes the $\nabla \cdot \boldsymbol{\sigma}$ term describing the forces within a fluid. At a macroscopic level, the forces exist due to interactions between fluid particles and change due to deformation of the fluid as it moves.

From the velocity field $\mathbf{u}$, we need to isolate pure deformation of the fluid from other characteristics of the flow. The figure below shows a rectangular fluid element under shear in the $x$-direction due to a uniform velocity gradient in the $y$-direction.

The velocity gradient $\nabla\mathbf{u}$ calculates *some* rate of shear of the rectangular element. The upper diagram shows shear in one direction, known as *parallel shear*.

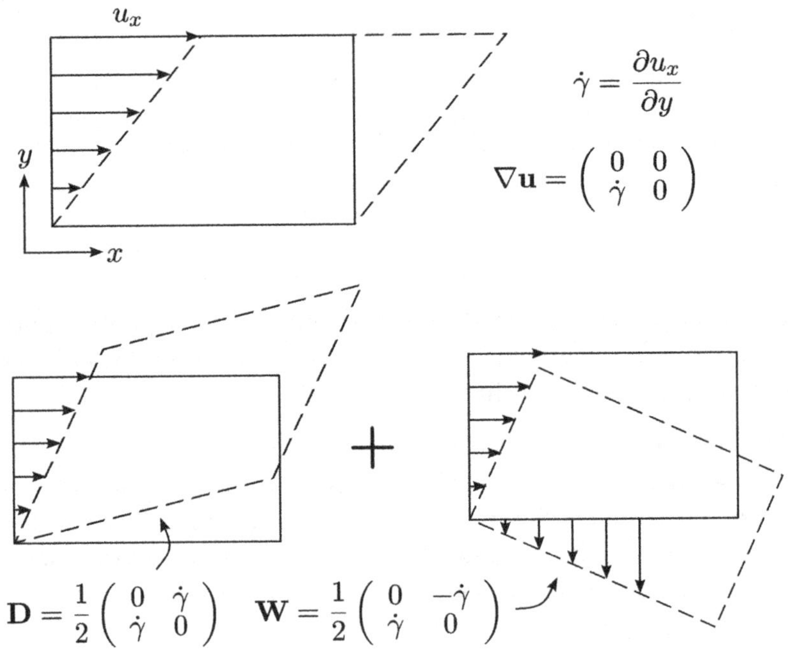

Parallel shear can be decomposed into pure shear (left) and a rotation (right). The rotational component is represented by the skew part of $\nabla\mathbf{u}$, see Eq. (2.25). Since a rotation does not involve deformation, it must be removed from any measure of it, which leaves the symmetric part of the $\nabla\mathbf{u}$ tensor.

Therefore, the rate of deformation tensor $\mathbf{D}$ and spin (rotational) tensor $\mathbf{W}$ are defined as

$$\mathbf{D} = \mathrm{sym}(\nabla\mathbf{u}) \quad \text{and} \quad \mathbf{W} = \mathrm{skw}(\nabla\mathbf{u}). \tag{2.33}$$

## Trace of a tensor

The trace of a tensor $\mathbf{T}$ is the sum of the diagonal components, producing a scalar

$$\mathrm{tr}(\mathbf{T}) = T_{xx} + T_{yy} + T_{zz}. \tag{2.34}$$

Where the tensor is $\nabla\mathbf{u}$, we conclude from Eq. (2.13) that

$$\mathrm{tr}(\nabla\mathbf{u}) = \frac{\partial u_x}{\partial x} + \frac{\partial u_y}{\partial y} + \frac{\partial u_z}{\partial z} = \nabla\bullet\mathbf{u}. \tag{2.35}$$

## Deviatoric and spherical tensors

A tensor can be decomposed into the sum of its deviatoric part and a spherical part as follows:

$$\mathbf{T} = \underbrace{\mathbf{T} - \frac{1}{3}(\mathrm{tr}\,\mathbf{T})\,\mathbf{I}}_{\textit{deviatoric}} + \underbrace{\frac{1}{3}(\mathrm{tr}\,\mathbf{T})\,\mathbf{I}}_{\textit{spherical}} = \mathrm{dev}\,\mathbf{T} + \mathrm{sph}\,\mathbf{T}, \tag{2.36}$$

where $\mathbf{I}$ is the identity tensor, see Eq. (2.30). The deviatoric part subtracts the mean of the trace from each diagonal component such that the resulting tensor is "trace-free", *i.e.* $\mathrm{tr}(\mathrm{dev}\,\mathbf{T}) = 0$.

## 2.11 Vorticity

Vorticity describes the tendency for a fluid to rotate locally, defined as

$$\boldsymbol{\omega} = \nabla \times \mathbf{u}\,, \tag{2.37}$$

where the $\nabla\times$ operator is the *curl derivative.* The curl of a vector $\mathbf{u}$ is evaluated in Cartesian co-ordinates by

$$\nabla \times \mathbf{u} = \left( \frac{\partial u_z}{\partial y} - \frac{\partial u_y}{\partial z}, \frac{\partial u_x}{\partial z} - \frac{\partial u_z}{\partial x}, \frac{\partial u_y}{\partial x} - \frac{\partial u_x}{\partial y} \right). \tag{2.38}$$

The local rotation which vorticity measures is not due to the flow following a curved path. In fact, vorticity is often associated with shear when streamlines may be straight and parallel, *i.e.* not curved at all.

Vorticity is difficult to picture under shear because the deformation masks the local rotation. By separating the deformation, as in the figure in Sec. 2.10, the local rotation is revealed.

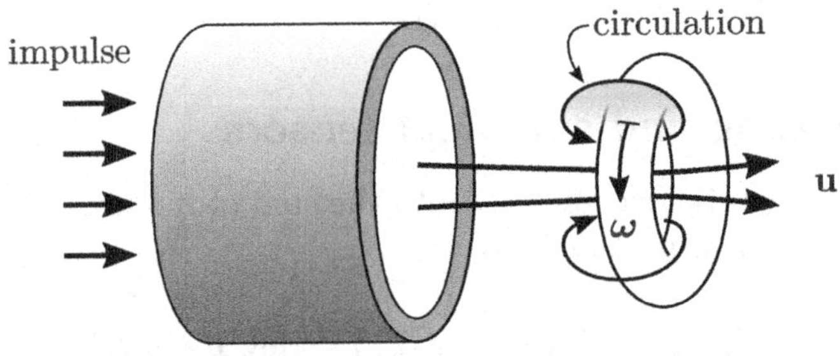


Vorticity $\boldsymbol{\omega}$ is often demonstrated by a vortex ring produced by an air "cannon" with smoke to visualise the flow. It reveals flow *circulation* around sections of the torus, from front to back. The vorticity vectors are normal to the planes of circulation, along the axis of the torus.

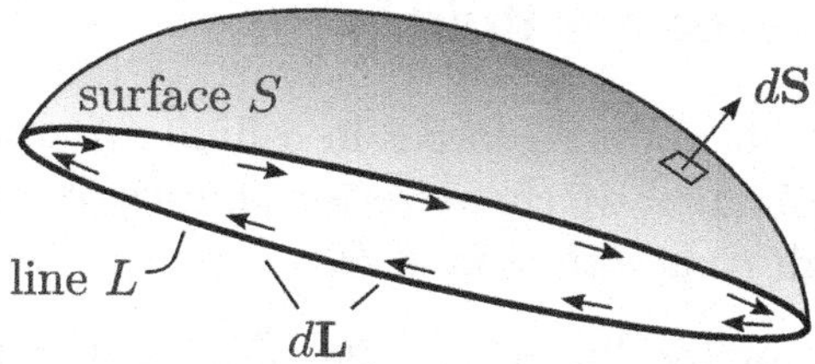


Stokes's theorem[†] relates *circulation* — the integral of $\mathbf{u}$ around a closed curved line, $L$ — to the integral of vorticity $\boldsymbol{\omega} = \nabla \times \mathbf{u}$ over a section of surface $S$ bounded by the curve, according to:

$$\int_S [d\mathbf{S} \cdot (\nabla \times \mathbf{u})] = \int_S (d\mathbf{S} \cdot \boldsymbol{\omega}) = \int_L (d\mathbf{L} \cdot \mathbf{u}). \tag{2.39}$$

In Eq. (2.39), $d\mathbf{L}$ is a vector representing a segment of the line $L$. As you stand on $L$ looking in the direction $d\mathbf{L}$, with your head in the direction $d\mathbf{S}$, $S$ is oriented to your left.

Vorticity is related to the spin tensor $\mathbf{W} = \text{skw}(\nabla \mathbf{u})$ in Eq. (2.33) by $\boldsymbol{\omega} \equiv 2\,{*}(\mathbf{W})$, where ${*}(\mathbf{T})$ is the Hodge dual operator which extracts components of a vector from a tensor $\mathbf{T}$ as shown below:

$$*\,\mathbf{T} = (T_{yz}, -T_{xz}, T_{xy})\,. \tag{2.40}$$

$$\mathbf{T} = \begin{pmatrix} T_{xx} & T_{xy} & T_{xz} \\ T_{yx} & T_{yy} & T_{yz} \\ T_{zx} & T_{zy} & T_{zz} \end{pmatrix}$$

The term "spin tensor" emphasises *local* rotation as opposed to the general path in a similar way that the spin on a ball is clearly distinguished from its trajectory, which curves due to gravity.

[†]posed by George Stokes as a examination question for the 1854 Smith's Prize for progress in mathematics and natural philosophy at the University of Cambridge, awarded to James Clerk Maxwell and Edward Routh.

## 2.12 Newtonian fluid

In Sec. 2.6 we introduced forces in a fluid. At the macroscopic scale of a continuum, we characterise a fluid's response to applied forces through *constitutive models.*

The Newtonian (or linear viscous) fluid is the most common constitutive model that represents the behaviour of many liquids and gases. It states that a fluid at rest (or uniform velocity) does not sustain shear stress $\boldsymbol{\tau}$; it can be expressed by

$$\boxed{\boldsymbol{\tau} = 2\mu \operatorname{dev} \mathbf{D} \quad \text{and} \quad \boldsymbol{\sigma} = \boldsymbol{\tau} - p\mathbf{I}} \tag{2.41}$$

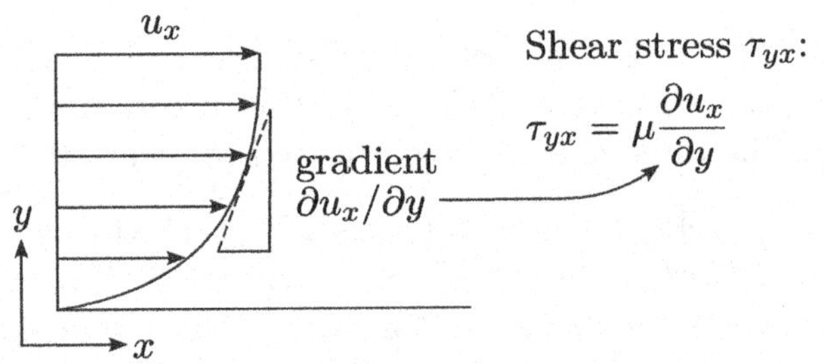


The model is a continuum representation† of Newton's law of viscosity‡ which states that shear stress $\tau_{yx}$ is proportional to velocity gradient by the dynamic viscosity $\mu$.

It is a specific case of the more general Stokesian fluid, defined as $\boldsymbol{\sigma} = f(\mathbf{D}) - p\mathbf{I}$, where $\mathbf{D}$ is deformation rate, Eq. (2.33). The Newtonian model assumes: (1) the fluid is isotropic, *i.e.* the value of $\mu$ is independent of the direction in which it is measured; (2) zero *bulk* viscosity associated with a change in volume o the fluid.

†attributed to Adhémar Jean Claude Barré de Saint-Venant, 1843, and George Gabriel Stokes, 1845; derived previously using molecular models by Claude-Louis Navier, *Mémoire sur les lois du mouvement des fluides*, 1822.

‡Isaac Newton, *Philosophiae Naturalis Principia Mathematica*, 1687.

Using $\mathbf{D}$ in Eq. (2.41) ensures that the shear stress is induced by deformation only. Taking the deviatoric part, $\text{dev}(\mathbf{D})$, ensures viscous stresses are not generated by volume changes, which are represented by $\nabla \bullet \mathbf{u}$.

This is due to the "dev" operator subtracting $\text{tr}(\mathbf{D})/3$ from *each* diagonal component of $\mathbf{D}$, giving a total of $\text{tr}(\mathbf{D})$ from all 3 diagonal components. From Eq. (2.35), $\text{tr}(\mathbf{D}) = \text{tr}(\nabla\mathbf{u}) = \nabla \bullet \mathbf{u}$.

## Pressure gradient

When we substitute Eq. (2.41) in $\nabla \bullet \sigma$ in Eq. (2.19), the pressure part is $-\nabla \bullet (p\mathbf{I})$. The term $\nabla \bullet (p\mathbf{I})$ is equivalent to gradient $\nabla p$.

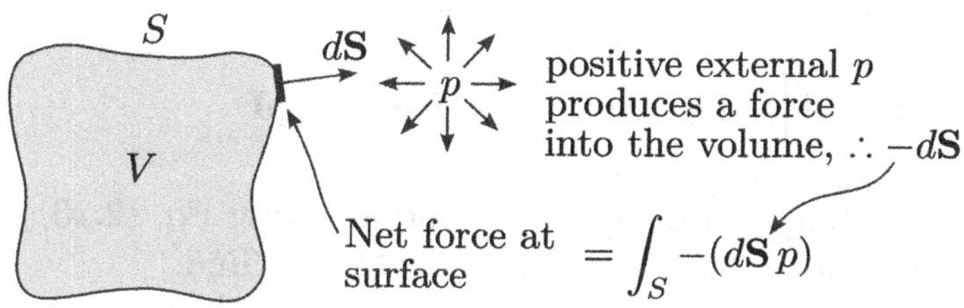


Like divergence of stress in Eq. (2.22), the gradient of pressure represents the pressure flux across the surface per unit volume as $\Delta V \to 0$, according to

$$\nabla p = \lim_{\Delta V \to 0} \frac{1}{\Delta V} \int_S (d\mathbf{S}\, p). \tag{2.42}$$

Since $\nabla \bullet (\mathbf{\Psi I}) \equiv \nabla \mathbf{\Psi}$ holds for any variable $\mathbf{\Psi}$, a gradient term is conservative, like divergence, and can be converted to a surface integral under an equivalent Gauss's Theorem

$$\int_S (d\mathbf{S}\, \mathbf{\Psi}) = \int_V \nabla \mathbf{\Psi}\, dV\,. \tag{2.43}$$

## 2.13 Incompressible flow

We can derive a set of equations for incompressible flow that includes:

- mass conservation, Eq. (2.8);
- momentum conservation, Eq. (2.19);
- the material derivative for $\mathbf{u}$, Eq. (2.26);
- the rate of deformation tensor, Eq. (2.33);
- the Newtonian fluid model, Eq. (2.41).

Combining Eq. (2.41) and Eq. (2.33) and using Eq. (2.36) for the deviatoric part of a tensor, gives an expression for stress:

$$\boldsymbol{\sigma} = \mu \left[ \nabla\mathbf{u} + \nabla\mathbf{u}^{\mathrm{T}} - \frac{2}{3}\,\mathrm{tr}(\nabla\mathbf{u})\mathbf{I} \right] - p\mathbf{I}. \tag{2.44}$$

Substituting $\boldsymbol{\sigma}$ into Eq. (2.19) and applying Eq. (2.26) gives an equation for momentum for a Newtonian fluid:

$$\begin{aligned} \frac{\partial \rho\mathbf{u}}{\partial t} + \nabla \bullet (\rho\mathbf{u}\mathbf{u}) - \nabla \bullet (\mu\nabla\mathbf{u}) - \nabla \bullet \left[\mu(\nabla\mathbf{u})^{\mathrm{T}}\right] \\ + \frac{2}{3}\mu\nabla(\nabla \bullet \mathbf{u}) = -\nabla p + \rho\mathbf{b}. \end{aligned} \tag{2.45}$$

The derivation of the equation above uses Eq. (2.35) and the identity $\nabla \bullet (\Psi\mathbf{I}) = \nabla\Psi$ on page 30.

An incompressible fluid exhibits $\rho$ = constant over time, *i.e.* for *moving volumes of fluid* $D\rho/Dt = 0$. Combining the material derivative Eq. (2.14) and mass conservation Eq. (2.8) gives $D\rho/Dt + \rho(\nabla \bullet \mathbf{u}) = 0$, which results in the *incompressibility condition*

$$\nabla \bullet \mathbf{u} = 0. \tag{2.46}$$

A *homogeneous*, incompressible material exhibits $\rho =$ constant uniformly throughout the *entire fluid*. With that assumption, Eq. (2.45) can be written as

$$\begin{aligned}\frac{\partial \mathbf{u}}{\partial t} + \nabla \bullet (\mathbf{u}\mathbf{u}) - \nabla \bullet (\nu \nabla \mathbf{u}) \\ - (\nabla \nu) \bullet (\nabla \mathbf{u}) = -\nabla p + \mathbf{b}.\end{aligned} \tag{2.47}$$

This is the momentum equation for a homogeneous, incompressible Newtonian fluid. It includes: *kinematic* viscosity $\nu = \mu/\rho$ in SI units of $\mathrm{m^2 s^{-1}}$; and $p$ represents *kinematic* pressure, *i.e.* divided by $\rho$, in SI units of $\mathrm{m^2 s^{-2}}$.

The identity $\nabla \bullet \left[\nu (\nabla \mathbf{u})^{\mathrm{T}}\right] \equiv \nu \nabla (\nabla \bullet \mathbf{u}) + (\nabla \nu) \bullet (\nabla \mathbf{u})$ and the incompressibility Eq. (2.46) yield the terms in Eq. (2.47).

## Pressure equation

Mass and momentum conservation, represented by Eq. (2.46) and Eq. (2.47) respectively, provide two equations — one scalar, one vector — for two fields, $p$ and $\mathbf{u}$. However, Eq. (2.46) cannot be solved in its own right since it provides only one equation for vector $\mathbf{u}$, containing 3 components.

A scalar equation, including both $p$ and $\mathbf{u}$, can be derived by taking *the divergence* of Eq. (2.47) and eliminating terms by substituting Eq. (2.46), noting that $\nabla \bullet (\nabla \bullet \nabla \mathbf{u}) \equiv \nabla^2 (\nabla \bullet \mathbf{u})$. For $\nu$ and $\mathbf{b}$ that are constant and uniform, the equation is

$$\nabla^2 p + \nabla \bullet [\nabla \bullet (\mathbf{u}\mathbf{u})] = 0. \tag{2.48}$$

This pressure equation can replace Eq. (2.46) to provide a pair of equations, with Eq. (2.47), for both $p$ and $\mathbf{u}$. Chapter 5 describes the algorithms used in the finite volume method which couple the two equations using a modified form of Eq. (2.48).

## 2.14 Diffusion

The momentum equation for a homogeneous, incompressible, Newtonian fluid is presented in Eq. (2.47). In the case of zero body force, $\mathbf{b} = \mathbf{0}$, and $\nu =$ constant, Eq. (2.47) becomes

$$\frac{\partial \mathbf{u}}{\partial t} + \nabla \bullet (\mathbf{u}\mathbf{u}) - \nabla \bullet (\nu \nabla \mathbf{u}) = -\nabla p. \tag{2.49}$$

This is a typical *transport equation* characterised by:

- the local rate of change $\partial \mathbf{u}/\partial t$, described in Sec. 2.5;
- advection of $\mathbf{u}$ by $\nabla \bullet (\mathbf{u}\mathbf{u})$, described in Sec. 2.8;
- diffusion of $\mathbf{u}$ by $\nabla \bullet (\nu \nabla \mathbf{u})$;
- a "source" due to $\nabla p$, described in Sec. 2.12.

Note that, since $\nu =$ constant, $\nabla \bullet (\nu \nabla \mathbf{u}) = \nu \nabla \bullet (\nabla \mathbf{u}) = \nu \nabla^2 \mathbf{u}$, where $\nabla^2 \equiv \nabla \bullet \nabla$ denotes the Laplace operator.

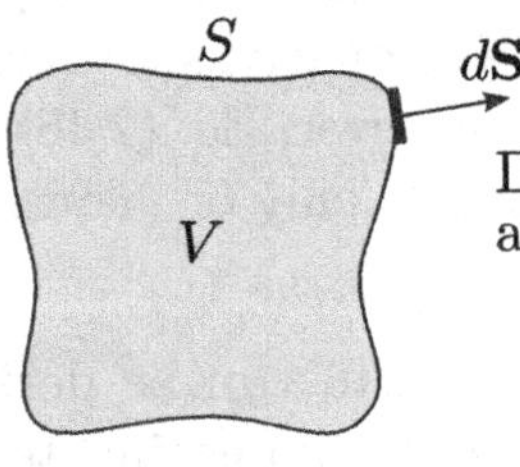


Diffusion at surface $= \displaystyle\int_S (d\mathbf{S} \bullet \nu \nabla \mathbf{u})$

$$= \int_S (dS\, \nu \nabla_{\mathbf{n}} \mathbf{u})$$

The $\nabla \bullet (\nu \nabla \mathbf{u})$ term is a special form of divergence in which the flux includes the *surface normal gradient* denoted by the operator $\nabla_{\mathbf{n}}$ where $\nabla_{\mathbf{n}} \boldsymbol{\Psi} \equiv \mathbf{n} \bullet \nabla \boldsymbol{\Psi}$.

The term models *diffusion* across the surface of the fluid element. Diffusion generally represents the transport of a fluid property — here, momentum — due to *fluctuating motions* that are not captured by the bulk motion that is represented by the *continuum* velocity $\mathbf{u}$.

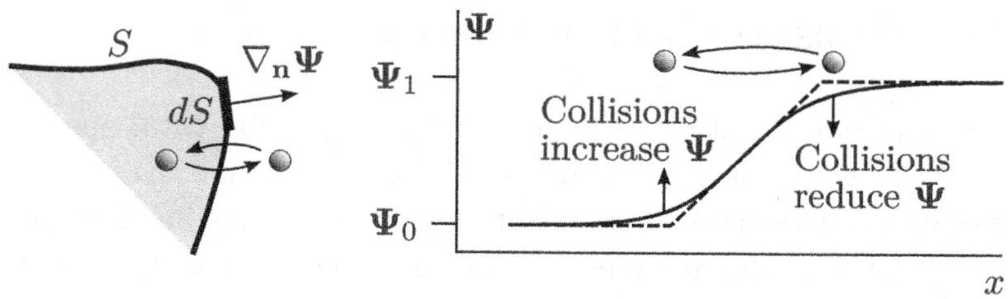


Fluctuations include any random motion of particle constituents of matter, *e.g.* molecules, and turbulent structures. Through these motions, particles can pass across a surface boundary, transporting property $\Psi$ through a gradient of $\Psi$ (above, left).

Particles carrying higher $\Psi$ move into regions of particles with lower $\Psi$ and vice versa. Through particle collisions, high values of $\Psi$ tend to reduce and low values increase (right).

## Laplacian

"Laplacian"† describes a term of the form $\nabla \cdot (\Gamma \nabla \Psi)$ where $\Gamma$ is a diffusivity coefficient.‡ A Laplacian term is *conservative* since all variables are to the right of a divergence, as described in Sec. 2.9. It is also *bounded* since it tends to decrease high values and increase low values as shown above.

The Laplacian represents a flux due to $\Gamma \nabla_{\mathbf{n}} \Psi$ across the surface, per unit volume, as $\Delta V \to 0$

$$\nabla \cdot (\Gamma \nabla \Psi) = \lim_{\Delta V \to 0} \frac{1}{\Delta V} \int_S (dS \, \Gamma \nabla_{\mathbf{n}} \Psi). \tag{2.50}$$

†After Pierre-Simon Laplace, *Théorie des attractions des sphéroïdes et de la figure des planètes*, 1785 but described in 1761 both by Leonhard Euler and Jean-Baptiste le Rond d'Alembert.

‡More precisely, "Laplacian" denotes a term without $\Gamma$, *i.e.* $\nabla \cdot (\nabla \Psi)$, represented as $\nabla^2 \Psi$ but we include $\Gamma$ here because it is generally needed.

## 2.15 Conservation of energy

The law of conservation of energy† states that the total energy of an isolated system remains constant over time. Energy is not created or destroyed, but is transformed from one form to another. If we consider mechanical and thermal energy, it can be expressed by

$$\boxed{\rho\frac{De}{Dt} + \rho\frac{DK}{Dt} = -\nabla \bullet \mathbf{q} + \rho r + \nabla \bullet (\boldsymbol{\sigma} \bullet \mathbf{u}) + \rho \mathbf{b} \bullet \mathbf{u}} \tag{2.51}$$

The equation describes the rate of change of mechanical and thermal energies for a volume of material of *fixed mass* of fluid particles. Thermal energy is described by the specific internal energy $e$, where *specific* denotes *per unit mass.* Mechanical energy is represented by specific kinetic energy $K \equiv |\mathbf{u}|^2/2$.

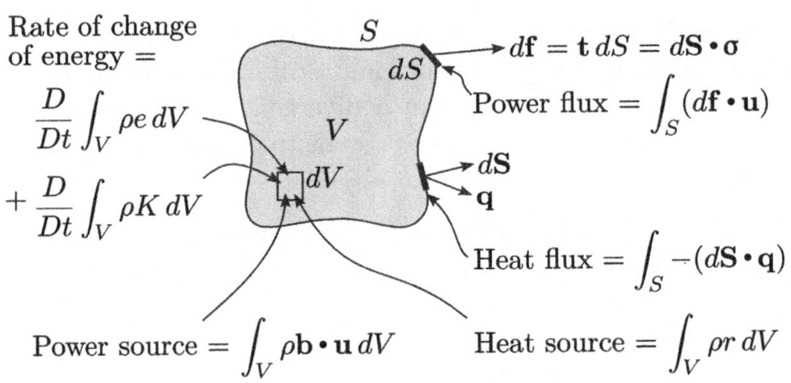


†from Hermann von Helmholtz, *Über die Erhaltung der Kraft*, 1847, who credited it equally to Julius von Mayer, 1842, and James Joule, 1843. "Joule's apparatus" is the best know demonstration in which the work of a falling weight is converted into heat in water. The first law of thermodynamics is the law of conservation of energy applied to a system.

The rate of change of these combined energies is due to the input of mechanical power and heat from both surface fluxes and internal sources.

Mechanical power is calculated by the inner product of force and $\mathbf{u}$. The power over a surface segment $dS$ is $d\mathbf{f} \bullet \mathbf{u} = (\mathbf{t}\, dS) \bullet \mathbf{u} = (d\mathbf{S} \bullet \boldsymbol{\sigma}) \bullet \mathbf{u}$, representing a change in strain energy. The internal power source is $\rho\mathbf{b} \bullet \mathbf{u}$, representing a change in potential energy due to general body forces.

Heat is provided by an internal source of strength $r$ per unit mass and a surface heat $\mathbf{q}$ per unit surface area. The surface heat flux relating to heat *input* takes a negative sign, $-\int_S (d\mathbf{S} \bullet \mathbf{q})$, since $\mathbf{q}$ follows the convention of pointing out of the volume.

Equation 2.51 is derived by equating integrals of rate of change of energy to the mechanical power and heat input and taking $D/Dt$ inside the volume integral since $\rho\, dV$ is constant in time, as follows:

$$\int_V \rho \left[\frac{De}{Dt} + \frac{DK}{Dt}\right] dV = \int_S [-d\mathbf{S} \bullet \mathbf{q} + (d\mathbf{S} \bullet \boldsymbol{\sigma}) \bullet \mathbf{u}] + \int_V [\rho r + \rho\mathbf{b} \bullet \mathbf{u}]\, dV. \tag{2.52}$$

Gauss's theorem converts the surface integrals of Eq. (2.52) into the following volume integrals:

$$\int_V [-\nabla \bullet \mathbf{q} + \nabla \bullet (\boldsymbol{\sigma} \bullet \mathbf{u})]\, dV. \tag{2.53}$$

Substituting Eq. (2.53) for the surface integrals into Eq. (2.52) and equating the integrand to 0, as described previously in Sec. 2.7, yields Eq. (2.51).

## 2.16 Temperature

In the conservation of energy Eq. (2.51), the mechanical kinetic energy, power flux and sources can be calculated from $\boldsymbol{\sigma}$, $\mathbf{u}$ and $\mathbf{b}$ from the momentum Eq. (2.19). Heat sources can contribute to $\rho r$, *e.g.* from thermal radiation, chemical reactions *etc.*

That leaves the heat flux term $\nabla \bullet \mathbf{q}$ which represents *conduction* of heat. It is commonly modelled by Fourier's law[†] which states $\mathbf{q}$ is proportional to the negative gradient of temperature $T$, *i.e.*

$$\boxed{\mathbf{q} = -\kappa \nabla T} \tag{2.54}$$

where the constant of proportionality is thermal conductivity $\kappa$.

### Temperature scale

The heat flux Eq. (2.54) requires temperature to be defined and measurable. Measurement requires a *scale.* Empirical scales correlate temperature with a measured physical property of a *working substance*, *e.g.* EMF at a junction of two metal alloys. Empirical scales have the drawbacks of: being dependent on the working substance; and, not actually defining temperature.

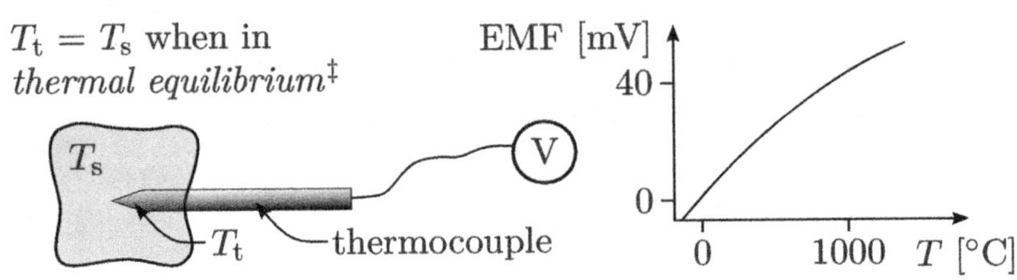


[†]Joseph Fourier, *Théorie analytique de la chaleur*, 1822.
[‡]in accordance with the zeroth law of thermodynamics.

Instead, the *thermodynamic scale* defines temperature as a measure of the average kinetic energy of random motions of particle constituents of matter. It provides an *absolute* measure of temperature that is independent of the choice of working substance and includes a zero point†. It must be measured in units with a zero point, such as the SI unit Kelvin, K.

Substitution of our model Eq. (2.54) into Eq. (2.51) yields the term $\nabla \bullet (\kappa \nabla T)$. It is logical that this is a Laplacian term since it represents diffusion which is associated with random motions of submicroscopic particles, as we we established in Sec. 2.14.

## Ideal gas

The behaviour of many gases under typical working conditions is captured by the ideal gas equation of state

$$p = \rho R T, \tag{2.55}$$

where the specific gas constant is $R = \mathbb{R}/M$. It is calculated from the Universal Gas Constant $\mathbb{R} \approx 8.314\,\mathrm{JK^{-1}mol^{-1}}$ in SI units, and the molar mass $M$ of the gas, with units of $\mathrm{kgmol^{-1}}$.

The ideal gas equation originates from classical thermodynamics as a combination of empirical laws‡. Later, it was derived from first principles from both statistical thermodynamics and kinetic theory, with temperature representing average kinetic energy.

The derivations assume that molecules have no volume, undergo purely elastic collisions and there are no inter-molecular forces.

A scale of temperature defined by the ideal gas equation of state is exactly equivalent to the thermodynamic temperature scale.

†in accordance with the third law of thermodynamics.

‡Benoît Clapeyron, *Mémoire sur la puissance motrice de la chaleur*, 1834.

## 2.17 Internal energy

The conservation of energy Eq. (2.51) introduces the internal energy using the specific quantity $e$, measured in $\mathrm{J\,kg^{-1}}$ in SI units. It represents the total molecular energy consisting of: kinetic energy associated with temperature; potential energy due to particle forces, both *within* particles as chemical bonds, and *between* particles, *e.g.* van der Waals forces.

The energies at different scales can be summarised as:

- bulk — kinetic $K$ due to bulk motion, potential due to forces $\boldsymbol{\sigma}$ and $\mathbf{b}$;
- molecular — kinetic characterised by $T$, potential due to bonds.

To understand how energy is transferred between these scales, we can derive an equation for internal energy from Eq. (2.51), by cancelling terms in mechanical energy formed by taking the inner product of Eq. (2.19) with $\mathbf{u}$.

$$\rho\frac{De}{Dt} + \underbrace{\rho\frac{DK}{Dt}}_{\cancel{\rho\frac{D\mathbf{u}}{Dt}}\cdot\mathbf{u}} = -\nabla\cdot\mathbf{q} + \rho r + \underbrace{\nabla\cdot(\boldsymbol{\sigma}\cdot\mathbf{u})}_{\cancel{(\nabla\cdot\boldsymbol{\sigma})}\cdot\mathbf{u} + \boldsymbol{\sigma}:\nabla\mathbf{u}} + \cancel{\rho\mathbf{b}\cdot\mathbf{u}}$$

$= 0$ by Eq. (2.19)

$$\boxed{\rho\frac{De}{Dt} = -\nabla\cdot\mathbf{q} + \rho r + \boldsymbol{\sigma}:\nabla\mathbf{u}} \qquad (2.56)$$

The identity $\nabla\cdot(\boldsymbol{\sigma}\cdot\mathbf{u}) \equiv (\nabla\cdot\boldsymbol{\sigma})\cdot\mathbf{u} + \boldsymbol{\sigma}:\nabla\mathbf{u}$ is the key element of the analysis. The $\boldsymbol{\sigma}:\nabla\mathbf{u}$ term in Eq. (2.56) must represent the contribution of mechanical power to the internal energy, *i.e.*

passing from bulk to molecular scale. Conversely, the $(\nabla \cdot \boldsymbol{\sigma}) \cdot \mathbf{u}$ term must contribute to mechanical energy.

Substituting $\boldsymbol{\sigma} = \boldsymbol{\tau} - p\mathbf{I}$ from Eq. (2.41), Eq. (2.56) becomes

$$\rho \frac{De}{Dt} = -\nabla \cdot \mathbf{q} + \rho r + \boldsymbol{\tau} : \nabla \mathbf{u} - p(\nabla \cdot \mathbf{u}). \tag{2.57}$$

The sign of $p(\nabla \cdot \mathbf{u})$ depends on $\nabla \cdot \mathbf{u}$, *i.e.* whether the fluid is expanding or contracting. Since the sign can change, it therefore represents a *recoverable* contribution to internal energy.

If we substitute the Newtonian model from Eq. (2.41)

$$\boldsymbol{\tau} : \nabla \mathbf{u} \equiv 2\mu(\operatorname{dev} \mathbf{D}) : \nabla \mathbf{u} \equiv 2\mu(\operatorname{dev} \mathbf{D}) : (\operatorname{dev} \mathbf{D}).^{\dagger} \tag{2.58}$$

The $\boldsymbol{\tau} : \nabla \mathbf{u}$ term is always positive since all components of $\operatorname{dev} \mathbf{D}$ are squared. Its contribution to $e$ is therefore *non-recoverable* and represents mechanical power that is dissipated as heat. In the majority of CFD analyses, the $\boldsymbol{\tau} : \nabla \mathbf{u}$ contribution is small and can be ignored.

## Double inner product of two tensors

The double inner product of two tensors, denoted by "$:$", is introduced in Eq. (2.56). It produces a scalar which is evaluated as the sum of the 9 products of the tensor components, eg:

$$\begin{aligned} \mathbf{T} : \mathbf{S} = \; & T_{xx}S_{xx} + T_{xy}S_{xy} + T_{xz}S_{xz} + \\ & T_{yx}S_{yx} + T_{yy}S_{yy} + T_{yz}S_{yz} + \\ & T_{zx}S_{zx} + T_{zy}S_{zy} + T_{zz}S_{zz} \end{aligned} \tag{2.59}$$

For a scalar $s$, $\mathbf{T} : s\mathbf{I} \equiv s\mathbf{I} : \mathbf{T} \equiv s \operatorname{tr}(\mathbf{T})$.

---

†This identity relies on the substitution of $\nabla \mathbf{u}$ by $\mathbf{D}$ which is equivalent due to symmetry of $\boldsymbol{\tau}$. $\mathbf{D}$ can then be replaced by $\operatorname{dev} \mathbf{D}$ since $(\operatorname{dev} \mathbf{D}) : (\operatorname{sph} \mathbf{D}) = 0$.

## 2.18 Heat capacity

When the heat conduction model of Eq. (2.54) is substituted into Eq. (2.51), we produce an equation with two variables $e$ and $T$. For example, starting with Eq. (2.56) for internal energy, and ignoring mechanical power and heat source terms, gives

$$\rho\frac{De}{Dt} - \nabla \bullet \kappa\nabla T = 0. \tag{2.60}$$

To be able to solve this equation, a relation between $e$ and $T$ is needed. *Heat capacity* provides this, describing an amount of heat required to produce a unit change in $T$. We use *specific* heat capacity, relating to a constant volume process, whose SI units are $\mathrm{J kg^{-1} K^{-1}}$, defined as

$$c_V = \left(\frac{\partial e}{\partial T}\right)_V. \tag{2.61}$$

Statistical mechanics provides a theorem known as *equipartition* which provides quantitative predictions of $c_V$.† It calculates that *any degree of freedom* (DoF), such as a component $v_x$ of particle translational velocity $\mathbf{v}$, which appears *quadratically in energy*, *i.e.* as $v_x^2$, contributes $R/2$ to $c_V$.

For monatomic gases, *e.g.* argon Ar, particle motion is translational only (until atoms dissociate into subatomic particles). With 3 translational DoFs, $c_V = 3R/2$ is constant.

A gas which obeys the ideal gas law with constant $c_V$ is known as *calorically perfect*.

---

†originating from heat capacity studies by Alexis Petit and Pierre Louis Dulong, *Recherches sur quelques points importants de la théorie de la chaleur*, 1819.

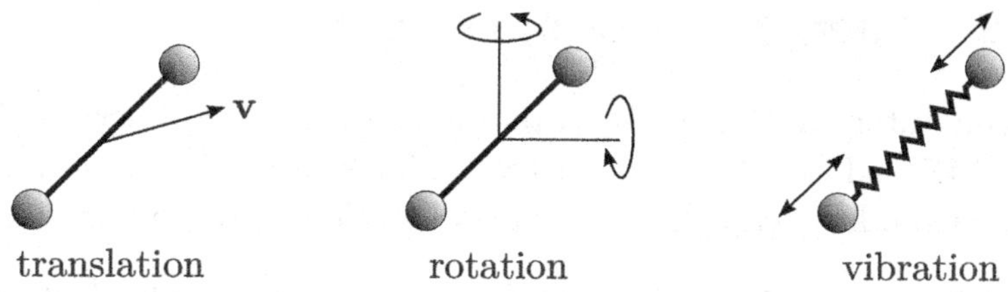


For diatomic gases, *e.g.* nitrogen $N_2$, molecular motion is translational (left) and rotational (centre) at lower temperature (in the gas state). At higher temperature, vibrational motion is excited along its bond (right). Heat capacity $c_V(T)$ for diatomic gases are then a function of temperature as shown below for $N_2$.

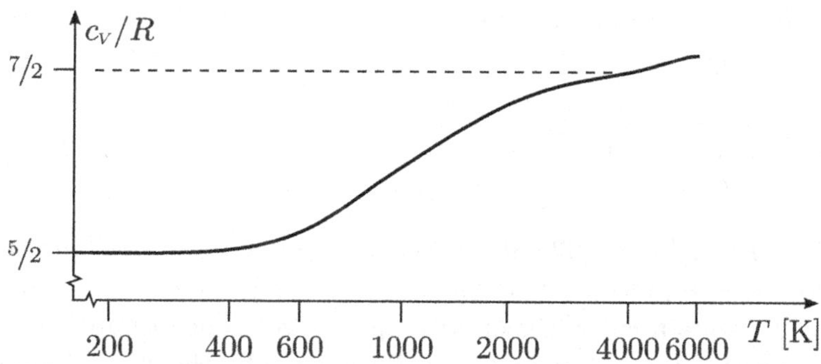


Rotational motion provides 2 DoFs (rotational energy about the axis is negligible), giving $c_V/R = 5/2$. Beyond 500 K, vibrational motion provides 2 additional DoFs (one kinetic, one potential), causing a gradual transition to $c_V/R = 7/2$. At high temperature, molecules dissociate and $c_V/R$ increases further.

A *thermally perfect* gas is an ideal gas with $c_V = c_V(T)$. Often $c_V$ can be treated as constant over a range of $T$, *e.g.* T $< 600$ K for $N_2$. Otherwise, accurate calculations require a suitable model of $c_V = c_V(T)$. An *imperfect* gas exhibits $c_V = c_V(T, p)$.

## 2.19 Energy and temperature

Specific internal energy $e$ and temperature $T$ were described in Sec. 2.17 and Sec. 2.16, respectively. They are related through the specific heat capacity $c_V$, defined by Eq. (2.61) in Sec. 2.18.

Analyses involving heat usually incorporate both $e$ and $T$ since:

- $T$ is the *measurable* quantity specified as initial and boundary data and whose data is required as part of the "results";
- $e$ is the *calculated* quantity solved in energy conservation, *e.g.* Eq. (2.51), but whose data is usually of no interest.

Conversion of values between $e$ and $T$ is therefore needed, and vice versa. Incorporating Eq. (2.61) into a definite integral for $e$, $\int_{e_{\text{ref}}}^{e} de$, gives

$$e = \int_{T_{\text{ref}}}^{T} c_V \, dT + e_{\text{ref}}. \tag{2.62}$$

The terms in Eq. (2.62) are illustrated below on a $c_V$–$T$ graph. Energy $e$ is represented by the area under the curve, in which $e_{\text{ref}}$ represents a reference energy up to a reference temperature $T_{\text{ref}}$, and the integral from $T_{\text{ref}}$ to $T$ is shown by the shaded area.

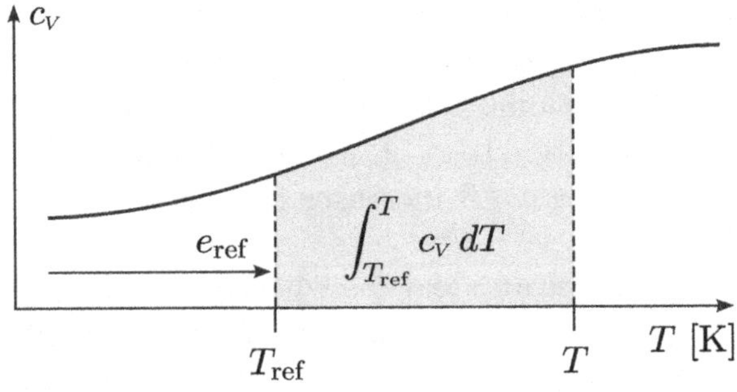


For applications that cover a reasonably narrow temperature

range, $c_V$ can be assumed constant. From Eq. (2.62), the $e$–$T$ relation becomes

$$e = c_V\,(T - T_{\text{ref}}) + e_{\text{ref}}. \tag{2.63}$$

Alternatively, $\int c_V\,dT$ can be integrated analytically by representing $c_V$ by a polynomial of order $n$ with coefficients $a_i$ ($i = 0, 1, \ldots, N$) fitted to measured $c_V$–$T$ data

$$c_V = \sum_{i=0}^{n} a_i T^i. \tag{2.64}$$

The values $e_{\text{ref}}$ and $T_{\text{ref}}$ ultimately add a constant component to $e$. Since Eq. (2.51) is concerned with *changes* in $e$ and the absolute values $e$ are usually of no interest, the values of $e_{\text{ref}}$ and $T_{\text{ref}}$ are often immaterial.

The $e_{\text{ref}}$ and $T_{\text{ref}}$ values become important when the composition of a fluid changes due to the mixing of constituent fluid species, *e.g.* $O_2$, $N_2$, or chemical reactions, *e.g.* with $CH_4$. Each fluid specie possesses a different $e_{\text{ref}}$ so any change to the specie concentrations will change $e$ of the overall fluid.

In those circumstances, $e_{\text{ref}}$ is commonly represented by the *heat of formation* per unit mass, $\Delta h_{\text{f}} = \Delta H_{\text{f}}/M$. The *standard heat of formation* $\Delta H_{\text{f}}$ is the change of enthalpy during the formation of 1 mole of a substance from its constituent elements at standard temperature $T^{\ominus} = 298.16\,\text{K}$. Measured heats of formation are available for numerous fluid species.†

If an analysis involves changes to fluid composition, it can then adopt $e_{\text{ref}} = \Delta h_{\text{f}}$ and $T_{\text{ref}} = T^{\ominus}$ for individual fluid species, to account for the change in $e$ due to changes in the concentrations of fluid species.

†Alexander Burcat and Branko Ruscic, *Third Millennium ideal gas and condensed phase thermochemical database for combustion*, 2005.

## 2.20 Natural convection

In Sec. 2.13, a set of equations — Eq. (2.47) and Eq. (2.48) — was derived for flow of an incompressible fluid. They are a sample set of equations for mass and momentum conservation that can be solved using methods described in this book.

The example set of equations can be extended to include energy conservation and associated models of heat conduction and heat capacity, described in Sec. 2.15 - 2.18.

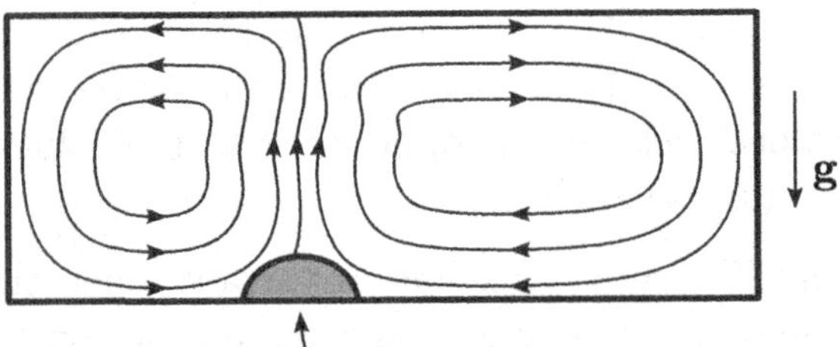


Heat source, $r$

The set of equations for mass, momentum and energy can be combined to simulate *natural convection*, *e.g.* for flow of air around a room. In natural convection, a non-uniform temperature causes density variations which generate associated forces due to gravity. Colder air is driven downwards and hot air rises, creating *buoyancy*. Small temperature variations, *e.g.* due to a heat source $r$, can cause buoyancy to be the dominant force.

A simple, approximate equation for $T$ can be derived, starting from internal energy conservation in the form of Eq. (2.57). The approximations of constant $\rho$ (with $\nabla \cdot \mathbf{u} = 0$) and zero viscosity reduce the stress/pressure work terms to zero.

Assuming $c_V$ = constant, we can apply Eq. (2.63), which reduces to substituting $e$ by $c_V T$ since derivatives of constants $c_V T_{\text{ref}}$ and

$e_{\text{ref}}$ are zero. Applying Fourier's law Eq. (2.54) leads to

$$\frac{\partial T}{\partial t} + \nabla \cdot (\mathbf{u}T) - \nabla \cdot (\alpha \nabla T) = r, \tag{2.65}$$

where thermal diffusivity $\alpha = \kappa/\rho c_v$ and $r$ becomes a thermal source in SI units of $\text{Ks}^{-1}$.

This is another example, similar to Eq. (2.49), of a transport equation containing a time derivative, advection, diffusion and a source of heat. Applying suitable boundary conditions, the equation can be solved for $T$.

## Buoyancy force

The effect of buoyancy can be simulated by setting the body force $\mathbf{b}$ in Eq. (2.47) for an incompressible, Newtonian fluid. While the assumption $\rho$ = constant is applied across all the governing equations generally, it cannot be applied to this force. Therefore we apply

$$\mathbf{b} = \frac{\rho(T, p)}{\rho_0}\mathbf{g}, \tag{2.66}$$

where $\rho_0$ is a density at a reference state, *e.g.* at the initial $T$ and $p$, and $\mathbf{g}$ is the acceleration due to gravity.

The density $\rho(T, p)$ is a function of $T$ and, optionally, $p$ provided by some equation of state, *e.g.* the ideal gas Eq. (2.55). The final momentum equation including this buoyancy force, and assuming $\nu$ = constant is:

$$\frac{\partial \mathbf{u}}{\partial t} + \nabla \cdot (\mathbf{u}\mathbf{u}) - \nabla \cdot (\nu \nabla \mathbf{u}) = -\nabla p + \frac{\rho(T, p)}{\rho_0}\mathbf{g} \tag{2.67}$$

The combined set of equations for mass, momentum and energy becomes Eq. (2.48) Eq. (2.67) and Eq. (2.65) respectively, which can be used to solve flows with natural convection.

## 2.21 Scale similarity

Scale similarity is the notion that, for two systems that are geometrically similar, the flow will follow the same path if the ratio of magnitude of forces acting on the fluid is the same at different points in the flow.

Flows at different scales can be compared using dimensionless variables. The momentum equation, Eq. (2.67), with advection, diffusion and gravity forces can be expressed in non-dimensionalised form by

$$\mathrm{Sr}\,\frac{\partial \hat{\mathbf{u}}}{\partial \hat{t}} + \hat{\nabla} \bullet (\hat{\mathbf{u}}\hat{\mathbf{u}}) - \left(\frac{1}{\mathrm{Re}}\right)\hat{\nabla}^2\hat{\mathbf{u}} = - \mathrm{Eu}\,\hat{\nabla}\hat{p} + \left(\frac{1}{\mathrm{Fr}^2}\right)\mathbf{n}_{\mathrm{g}}\hat{\rho}, \tag{2.68}$$

with the dimensionless numbers†:

- Strouhal number $\mathrm{Sr} = L/(\mathcal{T}U)$ — transient/steady inertia;
- Reynolds number $\mathrm{Re} = UL/\nu$ — inertia/viscous force;
- Euler number $\mathrm{Eu} = p_0/U^2$ — pressure force/inertia;
- (Froude number)$^2$ $\mathrm{Fr}^2 = U^2/gL$ — inertia/gravity force.

These dimensionless numbers include a characteristic length $L$, time $\mathcal{T}$, speed $U$ and pressure $p_0$. The ^ (hat) notation indicates dimensionless length, time, *etc.*, *e.g.* $\hat{\mathbf{x}} = \mathbf{x}/L$ and $\hat{t} = t/\mathcal{T}$ and corresponding dimensionless operators $\partial/\partial\hat{t} = \mathcal{T}(\partial/\partial t)$ and $\hat{\nabla} = L\nabla$.

Equation 2.68 assumes constant $\nu$ and splits gravitational acceleration $\mathbf{g} = g\mathbf{n}_{\mathrm{g}}$ into its magnitude $g$ and unit direction $\mathbf{n}_{\mathrm{g}}$; pressure, including $p_0$, is in kinematic units (divided by $\rho$).

† After Vincenc Strouhal (1850-1922), Osborne Reynolds (1842-1912), Leonhard Euler (1707-1783) and William Froude (1810-1879).

The dimensionless numbers provide a comparison of the magnitudes of different fluid forces. For example, Re represents the ratio of inertia force to viscous force and plays a pivotal role in turbulence modelling, introduced in Chapter 6.

Scale similarity applies also to other transported properties. For example, the energy equation, Eq. (2.65), ignoring heat sources $r$, can be expressed in non-dimensional form as

$$\mathrm{Sr}\,\frac{\partial \hat{T}}{\partial \hat{t}} + \hat{\nabla} \bullet (\hat{\mathbf{u}}\hat{T}) - \left(\frac{1}{\mathrm{Pe}}\right) \hat{\nabla}^2 \hat{T} = 0, \tag{2.69}$$

which includes the following additional dimensionless number:†

- Péclet number $\mathrm{Pe} = UL/\alpha$ — advection/diffusion of heat.

Again in Eq. (2.69), the ^ (hat) notation is applied to temperature $\hat{T}$ to indicate a dimensionless temperature, although it notably does not appear in a dimensionless number.

In fact, with the exception of momentum (which uses Re), Pe represents more generally the rates of advection and diffusion, as a ratio, for any transported quantity (here, it is heat).

Further numbers‡ define the ratios of diffusivities, *e.g.*:

- Prandtl number $\mathrm{Pr} = \nu/\alpha$ — viscosity/thermal diffusivity;
- Schmidt number $\mathrm{Sc} = \nu/D$ — viscosity/mass diffusivity.

where $D$ is mass diffusivity (not discussed in this book).

Dimensionless numbers can be multiplied and divided with one another to form further dimensionless numbers. For example, the Péclet number for heat transfer $\mathrm{Pe} = \mathrm{Re}\ \mathrm{Pr}$.

† After Jean Claude Eugène Péclet (1793-1857).

‡ After Ludwig Prandtl (1875-1953) and Ernst Heinrich Wilhelm Schmidt (1892-1975).

## 2.22 Region of influence

A CFD calculation is performed by solving partial differential equations, such *e.g.* for conservation of mass, momentum and energy, over a solution domain. It requires suitable boundary conditions, discussed in Chapter 4.

The solution is influenced by any change in a field value, *e.g.* $p$, $\mathbf{u}$, $T$, at some point in the domain, *e.g.* at an inlet boundary. The form of the equation determines the way in which these changes, or *perturbations*, propagate across the domain over time.

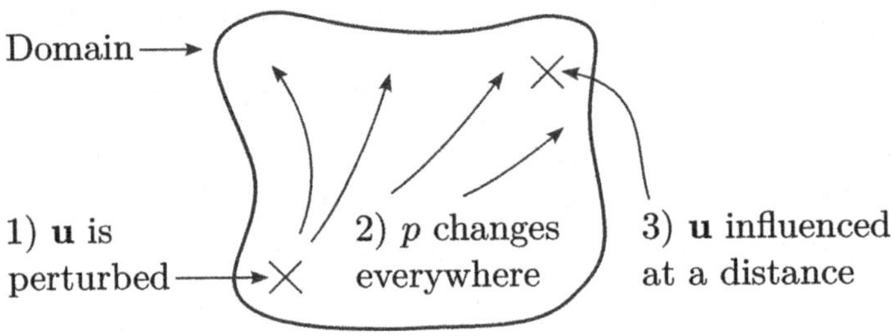


Momentum and mass conservation for an incompressible fluid can be represented by Eq. (2.67) and Eq. (2.48). The form of Eq. (2.48), including only a Laplacian derivative $\nabla^2 p$, ensures that $p$ is influenced instantaneously at all points in the domain by a perturbation in $\mathbf{u}$ at any point.

The resulting instantaneous change in $\nabla p$ then causes $\mathbf{u}$ to be redistributed everywhere by Eq. (2.67), with further short-range changes due to advection and diffusion (discussed next).

The outcome for an incompressible fluid is that $\mathbf{u}$ is influenced instantaneously everywhere in the domain by a perturbation at any point. In other words, the speed of sound $c$, corresponding to propagation of disturbances, is infinite.

## Advection-diffusion equations

Energy conservation can be represented by Eq. (2.65), in which perturbations propagate at a characteristic speed due to the advection and diffusion.

Advection propagates at speed $|\mathbf{u}|$, with a region of influence $\delta = |\mathbf{u}|t$ in the direction of flow. The relation can be obtained from scale similarity when $\nabla \cdot (\mathbf{u}T)$ and $\partial T/\partial t$ are similar in magnitude, *i.e.* $\mathrm{Sr} = 1$ in Eq. (2.69).

By the same argument, the region of influence for diffusion coincides with $\nabla \cdot (\alpha \nabla T)$ and $\partial T/\partial t$ being similar in magnitude, *i.e.* $\mathrm{Sr}\,\mathrm{Pe} = L^2/(\mathcal{T}U) = 1$. A diffusion "front" travels a distance according to $\delta \sim \sqrt{\alpha t}$ ($\sim$ means "of the order of magnitude").

The diffusion equation, $\partial \boldsymbol{\Psi}/\partial t = \alpha(\partial^2 \boldsymbol{\Psi}/\partial x^2)$, in one dimension ($x$) has a solution[†] $\boldsymbol{\Psi} = A \operatorname{erf}(x/2\sqrt{\alpha t}) + B$, where $A$, $B$ are constants and $\operatorname{erf}(\eta)$ is the error function below.

$$\operatorname{erf}(\eta) = \frac{2}{\sqrt{\pi}} \int_0^{\eta} \exp(-\eta^2)\, d\eta$$

| $\eta$ | 1 | 1.386 | 2 | $\infty$ |
|---|---|---|---|---|
| $\operatorname{erf}(\eta)$ | 0.843 | 0.950 | 0.995 | 1 |

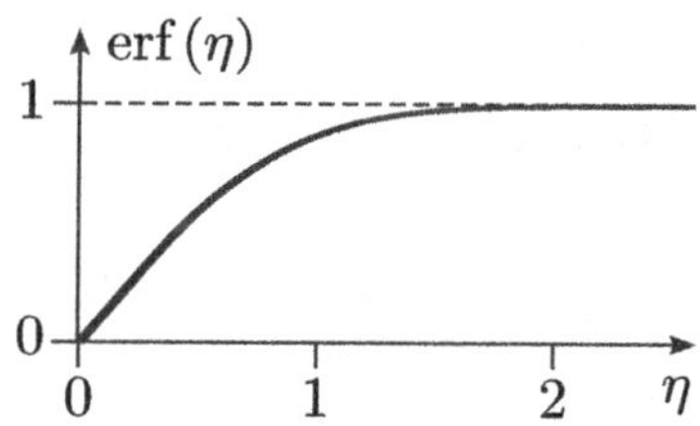


This means that for heat conduction problems, *e.g.* in solids, the distance travelled by the thermal front is $\delta = 2\eta\sqrt{\alpha t}$, consistent with similarity arguments above.

The coefficient $2\eta$ depends on where on the $\operatorname{erf}(\eta)$ curve the front is located. One option is $\eta \approx 2$, where the solution is within 0.5% of the asymptote of 1.

[†]Horatio Carslaw and John Jaeger, *Conduction of Heat in Solids.*

## 2.23 Summary of equations

- **Force** $d\mathbf{f}$ at a surface:

| | |
|---|---|
| $d\mathbf{f} = d\mathbf{S}\,p$ | pressure, Sec. 2.1 |
| $d\mathbf{f} = \mathbf{t}\,dS$ | traction, Sec. 2.6 |
| $d\mathbf{f} = d\mathbf{S} \bullet \boldsymbol{\sigma}$ | stress, Sec. 2.6 |

- **Rate of deformation**

$$\mathbf{D} = \mathrm{sym}(\nabla\mathbf{u}) \quad \text{Sec. 2.10}$$

## Conservation laws

- **Mass**, Sec. 2.4

$$\frac{\partial \rho}{\partial t} + \nabla \bullet (\rho\mathbf{u}) = 0$$

- **Momentum**, Sec. 2.7

$$\rho \frac{D\mathbf{u}}{Dt} = \nabla \bullet \boldsymbol{\sigma} + \rho\mathbf{b}$$

- **Energy**, Sec. 2.15

$$\rho \frac{De}{Dt} + \rho \frac{DK}{Dt} = -\nabla \bullet \mathbf{q} + \rho r + \nabla \bullet (\boldsymbol{\sigma} \bullet \mathbf{u}) + \rho\mathbf{b} \bullet \mathbf{u}$$

## Constitutive models

- **Newtonian fluid**, Sec. 2.12

$$\boldsymbol{\tau} = 2\mu\,\mathrm{dev}\,\mathbf{D} \quad \text{and} \quad \boldsymbol{\sigma} = \boldsymbol{\tau} - p\mathbf{I}$$

- **Fourier's law**, Sec. 2.16

$$\mathbf{q} = -\kappa \nabla T$$

- **Ideal gas**, Sec. 2.16

$$p = \rho R T$$

- **Specific heat capacity**, Sec. 2.18

$$c_V = \left(\frac{\partial e}{\partial T}\right)_V$$

- **Energy-temperature relation**, Sec. 2.19

$$e = \int_{T_{\text{ref}}}^{T} c_V \, dT + e_{\text{ref}}.$$

## Derivatives

- **Material time derivative**, Sec. 2.5

$$\frac{D\boldsymbol{\Psi}}{Dt} = \frac{\partial \boldsymbol{\Psi}}{\partial t} + \mathbf{u} \cdot \nabla \boldsymbol{\Psi}$$

- **Divergence**, Sec. 2.4, $\boldsymbol{\Psi}$ is vector, tensor

$$\nabla \cdot \boldsymbol{\Psi} = \lim_{\Delta V \to 0} \frac{1}{\Delta V} \int_S (d\mathbf{S} \cdot \boldsymbol{\Psi})$$

- **Gradient**, Sec. 2.12

$$\nabla \boldsymbol{\Psi} = \lim_{\Delta V \to 0} \frac{1}{\Delta V} \int_S (d\mathbf{S} \, \boldsymbol{\Psi})$$

- **Laplacian**, Sec. 2.14, $\nabla_{\mathbf{n}} \equiv \mathbf{n} \cdot \nabla$

$$\nabla \cdot (\Gamma \nabla \boldsymbol{\Psi}) = \lim_{\Delta V \to 0} \frac{1}{\Delta V} \int_S (dS \, \Gamma \nabla_{\mathbf{n}} \boldsymbol{\Psi})$$

## 2.24 Summary of tensor algebra

Below is tensor algebra applied to Cartesian $(x, y, z)$ co-ordinates using: scalar $s$; vectors $\mathbf{a}$, $\mathbf{b}$; tensors $\mathbf{T}$, $\mathbf{S}$, $\mathbf{Y}$.

### Products

- **Inner product of two vectors**, Sec. 2.3

$$s = \mathbf{a} \bullet \mathbf{b} = a_x b_x + a_y b_y + a_z b_z$$

- **Outer product of two vectors**, Sec. 2.8

$$\mathbf{T} = \mathbf{ab} = \begin{pmatrix} a_x b_x & a_x b_y & a_x b_z \\ a_y b_x & a_y b_y & a_y b_z \\ a_z b_x & a_z b_y & a_z b_z \end{pmatrix}$$

- **Inner product of vector and tensor**, Sec. 2.6

$$\mathbf{b} = \mathbf{a} \bullet \mathbf{T} = \begin{pmatrix} a_x T_{xx} + a_y T_{yx} + a_z T_{zx} \\ a_x T_{xy} + a_y T_{yy} + a_z T_{zy} \\ a_x T_{xz} + a_y T_{yz} + a_z T_{zz} \end{pmatrix}$$

- **Inner product of two tensors**, Sec. 2.8

$$\mathbf{Y} = \mathbf{T} \bullet \mathbf{S} \quad Y_{ij} = T_{ix} S_{xj} + T_{iy} S_{yj} + T_{iz} S_{zj} \quad (i, j = x, y, z)$$

- **Double inner product of two tensors**, Sec. 2.17

$$\begin{aligned} s = \mathbf{T} \colon \mathbf{S} = \; & T_{xx} S_{xx} + T_{xy} S_{xy} + T_{xz} S_{xz} + \\ & T_{yx} S_{yx} + T_{yy} S_{yy} + T_{yz} S_{yz} + \\ & T_{zx} S_{zx} + T_{zy} S_{zy} + T_{zz} S_{zz} \end{aligned}$$

- **Cross product of two vectors**, first used in Sec. 3.3, produces a vector with components

$$\mathbf{a} \times \mathbf{b} = (a_y b_z - a_z b_y, a_z b_x - a_x b_z, a_x b_y - a_y b_x) \quad (2.70)$$

## Tensors and operations

- **Transpose of a tensor**, Sec. 2.7

$$\mathbf{T} = \begin{pmatrix} T_{xx} & T_{xy} & T_{xz} \\ T_{yx} & T_{yy} & T_{yz} \\ T_{zx} & T_{zy} & T_{zz} \end{pmatrix} \qquad \mathbf{T}^{\mathrm{T}} = \begin{pmatrix} T_{xx} & T_{yx} & T_{zx} \\ T_{xy} & T_{yy} & T_{zy} \\ T_{xz} & T_{yz} & T_{zz} \end{pmatrix}$$

- **Symmetric and skew tensors**, Sec. 2.7

$$\mathbf{T} = \underbrace{\frac{1}{2}(\mathbf{T} + \mathbf{T}^{\mathrm{T}})}_{symmetric} + \underbrace{\frac{1}{2}(\mathbf{T} - \mathbf{T}^{\mathrm{T}})}_{skew} = \operatorname{sym}\mathbf{T} + \operatorname{skw}\mathbf{T}$$

- **Trace of a tensor**, Sec. 2.10

$$\operatorname{tr}(\mathbf{T}) = T_{xx} + T_{yy} + T_{zz}$$

- **Identity tensor**, Sec. 2.8

$$\mathbf{I} = \begin{pmatrix} 1 & 0 & 0 \\ 0 & 1 & 0 \\ 0 & 0 & 1 \end{pmatrix}$$

$$\mathbf{I} \bullet \mathbf{T} \equiv \mathbf{T} \bullet \mathbf{I} \equiv \mathbf{T}$$

$$\mathbf{T} \mathbin{\mathbf{:}} s\mathbf{I} \equiv s \operatorname{tr}(\mathbf{T})$$

- **Deviatoric and spherical tensors**, Sec. 2.10

$$\mathbf{T} = \underbrace{\mathbf{T} - \frac{1}{3}(\operatorname{tr}\mathbf{T})\,\mathbf{I}}_{deviatoric} + \underbrace{\frac{1}{3}(\operatorname{tr}\mathbf{T})\,\mathbf{I}}_{spherical} = \operatorname{dev}\mathbf{T} + \operatorname{sph}\mathbf{T}$$

## 2.25 Vector identities

Several identities are listed below which can be verified under the assumption that the relevant derivatives exist and are continuous. The identities are expressed for: scalars $s$, $r$; vectors $\mathbf{a}$, $\mathbf{b}$, $\mathbf{c}$; tensor $\mathbf{T}$.

### Algebraic operations

$$\mathbf{a} \times \mathbf{b} \equiv -(\mathbf{b} \times \mathbf{a}) \tag{2.71a}$$

$$\mathbf{a} \times \mathbf{b} \bullet \mathbf{c} \equiv \mathbf{a} \bullet \mathbf{b} \times \mathbf{c} \tag{2.71b}$$

$$\mathbf{a} \times (\mathbf{b} \times \mathbf{c}) \equiv (\mathbf{a} \bullet \mathbf{c})\mathbf{b} - (\mathbf{a} \bullet \mathbf{b})\mathbf{c} \tag{2.71c}$$

$$(\mathbf{a} \times \mathbf{b}) \times \mathbf{c} \equiv (\mathbf{a} \bullet \mathbf{c})\mathbf{b} - (\mathbf{b} \bullet \mathbf{c})\mathbf{a} \tag{2.71d}$$

$$\mathbf{a} \bullet \mathbf{T} \equiv \mathbf{T}^{\mathrm{T}} \bullet \mathbf{a} \tag{2.71e}$$

### Gradient derivatives

$$\nabla(sr) \equiv s\nabla r + r\nabla s \tag{2.72a}$$

$$\nabla(s/r) \equiv (r\nabla s - s\nabla r)/r^2 \tag{2.72b}$$

$$\nabla(s\mathbf{a}) \equiv (\nabla s)\,\mathbf{a} + s\nabla\mathbf{a} \tag{2.72c}$$

$$\nabla(\mathbf{a} \bullet \mathbf{b}) \equiv \mathbf{a} \bullet \nabla\mathbf{b} + \mathbf{b} \bullet \nabla\mathbf{a} + \mathbf{a} \times (\nabla \times \mathbf{b}) + \mathbf{b} \times (\nabla \times \mathbf{a}) \tag{2.72d}$$

### Divergence derivatives

$$\nabla \bullet (\nabla \times \mathbf{a}) \equiv 0 \tag{2.73a}$$

$$\nabla \bullet (s\mathbf{a}) \equiv (\nabla s) \bullet \mathbf{a} + s\nabla \bullet \mathbf{a} \tag{2.73b}$$

$$\nabla \bullet (\mathbf{a}/s) \equiv [\,s\nabla \bullet \mathbf{a} - (\nabla s) \bullet \mathbf{a}\,]/s^2 \tag{2.73c}$$

$$\nabla \bullet (\mathbf{a} \times \mathbf{b}) \equiv (\nabla \times \mathbf{a}) \bullet \mathbf{b} - \mathbf{a} \bullet (\nabla \times \mathbf{b}) \tag{2.73d}$$

$$\nabla \bullet (\mathbf{ab}) \equiv \mathbf{a} \bullet \nabla\mathbf{b} + \mathbf{b}(\nabla \bullet \mathbf{a}) \tag{2.73e}$$

$$\nabla \bullet (\mathbf{T} \bullet \mathbf{a}) \equiv (\nabla \bullet \mathbf{T}) \bullet \mathbf{a} + \mathbf{T} : (\nabla\mathbf{a}) \tag{2.73f}$$

## Laplacian derivatives

$$\nabla \bullet (s\nabla \mathbf{a}) \equiv s\nabla^2 \mathbf{a} + (\nabla s) \bullet (\nabla \mathbf{a}) \tag{2.74a}$$

$$\nabla^2 (sr) \equiv s\nabla^2 r + 2(\nabla s) \bullet (\nabla r) + r\nabla^2 s \tag{2.74b}$$

$$\nabla^2 (s\mathbf{a}) \equiv s\nabla^2 \mathbf{a} + 2(\nabla s) \bullet (\nabla \mathbf{a}) + \mathbf{a}\nabla^2 s \tag{2.74c}$$

$$\nabla^2 (\nabla s) \equiv \nabla(\nabla^2 s) \tag{2.74d}$$

$$\nabla^2 (\nabla \bullet \mathbf{a}) \equiv \nabla \bullet (\nabla^2 \mathbf{a}) \tag{2.74e}$$

$$\nabla^2 (\nabla \times \mathbf{a}) \equiv \nabla \times (\nabla^2 \mathbf{a}) \tag{2.74f}$$

$$\begin{aligned} \nabla^2 (\mathbf{a} \bullet \mathbf{b}) \equiv\; & \mathbf{a} \bullet (\nabla^2 \mathbf{b}) - \mathbf{b} \bullet (\nabla^2 \mathbf{a}) \\ & + 2\nabla \bullet [\mathbf{b} \bullet \nabla \mathbf{a} + \mathbf{b} \times (\nabla \times \mathbf{a})] \end{aligned} \tag{2.74g}$$

## Curl derivatives

$$\nabla \times (\nabla s) \equiv \mathbf{0} \tag{2.75a}$$

$$\nabla \times (s\mathbf{a}) \equiv \nabla s \times \mathbf{a} + s(\nabla \times \mathbf{a}) \tag{2.75b}$$

$$\nabla \times (\mathbf{a}/s) \equiv [\, s(\nabla \times \mathbf{a}) - \nabla s \times \mathbf{a} \,] / s^2 \tag{2.75c}$$

$$\nabla \times (s\nabla r) \equiv (\nabla s) \times (\nabla r) \tag{2.75d}$$

$$\begin{aligned} \nabla \times (\mathbf{a} \times \mathbf{b}) \equiv\; & \mathbf{a}(\nabla \bullet \mathbf{b}) - \mathbf{b}(\nabla \bullet \mathbf{a}) \\ & + \mathbf{b} \bullet \nabla \mathbf{a} - \mathbf{a} \bullet \nabla \mathbf{b} \end{aligned} \tag{2.75e}$$

$$\nabla \times (\nabla \times \mathbf{a}) \equiv \nabla(\nabla \bullet \mathbf{a}) - \nabla^2 \mathbf{a} \tag{2.75f}$$

$$(\nabla \times \mathbf{a}) \times \mathbf{a} \equiv \mathbf{a} \bullet \nabla \mathbf{a} - \frac{1}{2}\nabla(\mathbf{a} \bullet \mathbf{a}) \tag{2.75g}$$

# Chapter 3

# Numerical Method

*"We were talking about the space between us all."*

The Beatles, *Within You, Without You* (1967).

Chapter 2 described:

- the main governing equations of fluid dynamics, namely conservation laws of mass, momentum and energy;
- standard models for diffusion of momentum and heat, namely Newton's law of viscosity and Fourier's law of conduction;
- the ideal gas equation of state and specific heat capacity;
- tensor algebra used in the equations and the derivatives of time, divergence, gradient and Laplacian.

Sets of governing equations and models are combined to simulate different flow problems. Numerical methods are used in CFD to solve the equations since closed-form, analytical solutions only exist for very simple geometries and flow regimes.

The numerical methods represent *continuous* physical entities by equivalent *discrete* entities, *i.e.*:

- *time* is split into intervals of duration $\Delta t$;
- *space* becomes cells of a mesh — the solution domain;

- *fields*, *e.g.* **u**, become discrete values, *e.g.* one per cell;
- *partial differential equations*, *e.g.* for momentum, become a set of linear equations.

There are numerous discretisation methods to create the set of linear equations, but this book specifically examines the finite volume method.

The sets of equations can be solved using a variety of algorithms, with finite volume discretisation lending itself towards iterative methods.

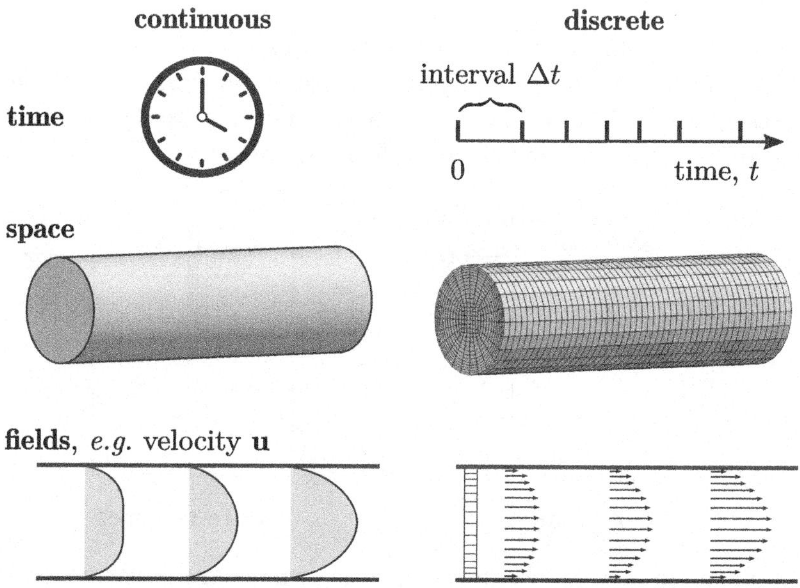


**equations**, *e.g.* momentum

$$\frac{\partial \mathbf{u}}{\partial t} + \nabla \cdot (\mathbf{u}\mathbf{u}) - \nabla \cdot (\nu \nabla \mathbf{u}) = -\nabla p \qquad \begin{bmatrix} \star & \star & & \star \\ \star & \star & \star & \\ & \star & \star & \star \\ \star & & \star & \star \end{bmatrix} \begin{bmatrix} u_x \end{bmatrix} = \begin{bmatrix} \star \\ \star \\ \star \\ \star \end{bmatrix}$$

## 3.1 The finite volume concept

There are many numerical methods which can be used to solve the partial differential equations encountered in fluid dynamics. No single method is unequivocally better than others. Rather, the efficacy of each method depends on aspects of the case being simulated, *e.g.* its size, the required level of accuracy and the characteristics and complexity of the equations being solved.

For a method to be useful, it needs to be programmed into software. Complex programs require good design and design relies on good concepts. So inevitably the concepts behind any method are as important as the details contained within.

The finite volume method adopts the idea of control volumes used to create models of physical systems. A control volume represents a region of space, which is generally fixed, enclosed by a surface through which fluid flows in and out.

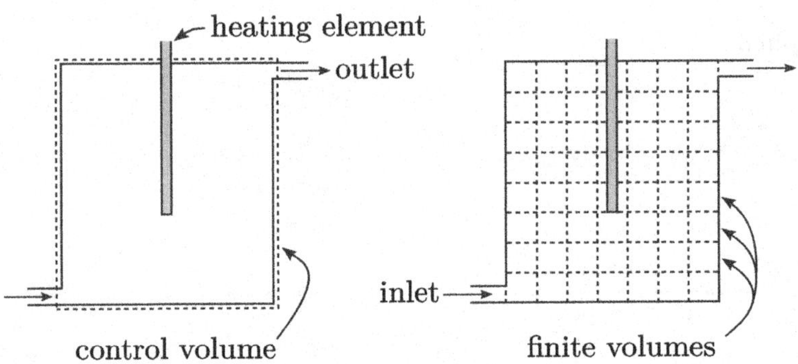


It applies conservation equations, *e.g.* of mass, momentum and energy, by balancing fluxes, due to inflow and outflow at the bounding surface, with additional sources within the volume. Rather than using a single control volume to describe an entire physical system, *e.g.* a heating tank, the finite volume method splits the system domain, *i.e.* the tank, into multiple connected finite volumes. Conservation equations are applied to each volume, ensuring that the fluxes of mass, momentum and heat across surfaces are consistent between the volumes they connect.

The perceived wisdom is that the finite volume method was first introduced in the early 1970's. But the PhD thesis of Runchal[†] from 1969 describes a method which is clearly finite volume. He also published the figure reproduced below,[‡] which displays the computational mesh as a set of connected control volumes.

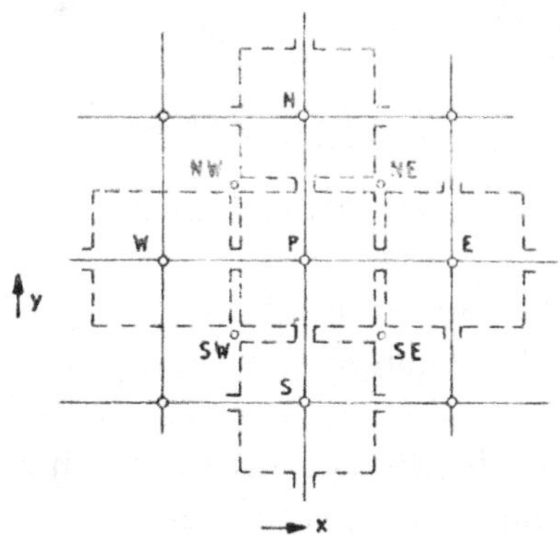


Original figure illustrating Spalding's "tanks and tubes". Each node is surrounded by a control volume (the tank). Control volumes are connected by mesh edges (the tubes).

Runchal credits the idea to his PhD supervisor Brian Spalding who provided an analogy of tanks connected by tubes in 1967. It captures the essence of finite volume, which uses a mesh to define a physical system of control volumes. Its emphasis is on calculating fluxes between volumes and ensuring conservation.

By contrast, other methods, *e.g.* finite element, use the mesh to construct mathematical functions to calculate distributions of properties. The finite volume method does not do that.

[†]Akshai Runchal, *Transfer processes in steady two-dimensional separated flows*, PhD Thesis, 1969.

[‡]Akshai Runchal, Brian Spalding and Micha Wolfshtein, *Numerical solution of the elliptic equations for transport of vorticity, heat, and matter in two-dimensional flow*, 1969.

## 3.2 Computational mesh

A CFD simulation begins with a *solution domain* which specifies a region of space of a particular geometric shape, in which fluid dynamics equations are solved. For example, when simulating flow along a pipe, the solution domain would be some region of space occupied by the fluid within the pipe.

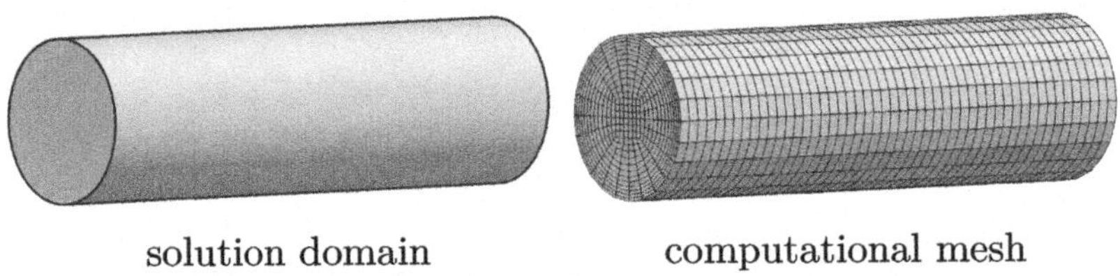


The process of *mesh generation* subdivides the solution domain into a mesh of small volumes, or *cells*. Computer programs are used to create the cells according to some user specification. We discuss neither mesh generation software nor any underlying methods they use. Rather, we define the structure and properties of a valid mesh in this chapter and some aspects of the design of optimal meshes in Chapter 8.

### Polyhedral mesh

Meshes in modern CFD software, that use the finite volume method, can contain cells of any irregular polyhedral shape. A cell can consist of an unlimited number of faces ($\geq 4$) and each face can have an unlimited number of edges ($\geq 3$).

Cells are *contiguous*, *i.e.* the faces of a given cell are common to its neighbouring cells (unless they form the boundary of the solution domain). There is no restriction on the alignment of cells with co-ordinate axes.

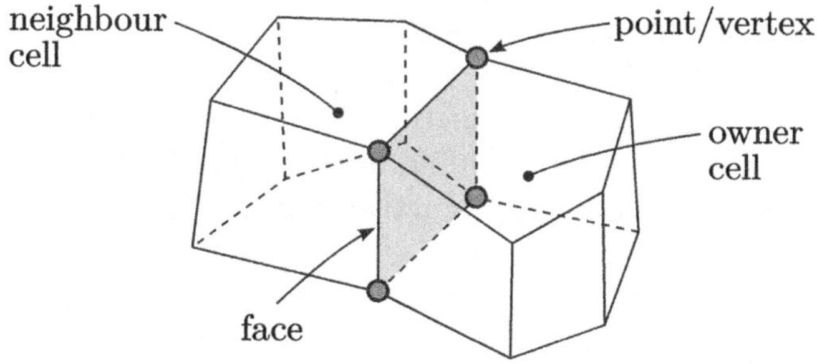


The figure above shows two cells connected by a common face displayed in grey. Every face is defined by a list of vertices that follow edges in sequence around the face.

Each face inside the mesh is common to two cells, one known as the *owner* cell, the other the *neighbour* cell. At the domain boundary, each face is connected to one owner cell. By assigning a unique index to each cell, the connectivity between cells can represented by storing the indices of the owner and neighbour cell of each face.

The cell faces that describe the domain boundary are separated into groups, each with a unique name. Each named group of boundary faces, known as a *patch*, identifies a specific region of the domain boundary, to enable us to apply specific boundary conditions to that region when running a CFD simulation.

A mesh is therefore defined by the following:

- a list of every vertex used to define all faces;
- a list of faces, each defined by a sequence of vertex indices;
- a list of owner cell indices associated with the faces;
- a list of neighbour cell indices associated with faces;
- grouping boundary faces into a set of patches.

## 3.3 Finite volume mesh

The finite volume numerical method is closely associated with the concept of surface and volume integrals used in Chapter 2. Below, we extract the main geometric elements from the figures used for the conservation laws.

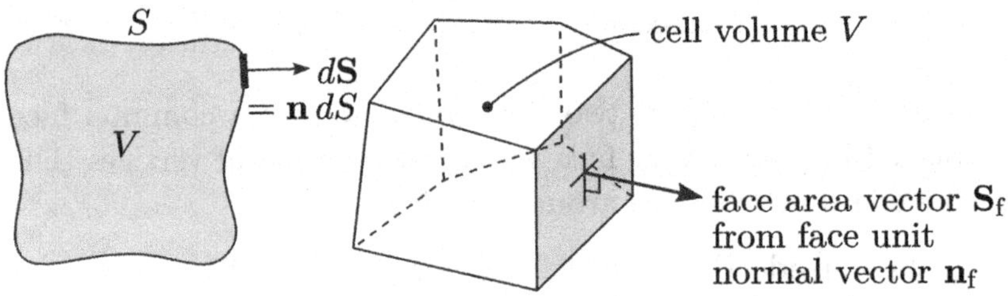


The integrals use volumes $dV$ and area vectors $d\mathbf{S} = \mathbf{n}\, dS$. The numerical method uses equivalent discrete quantities for cells and faces:

- cell volume $V$;
- face area vector $\mathbf{S}_{\mathrm{f}}$, with area magnitude $|\mathbf{S}_{\mathrm{f}}|$;
- face unit normal vector $\mathbf{n}_{\mathrm{f}} = \mathbf{S}_{\mathrm{f}}/|\mathbf{S}_{\mathrm{f}}|$.

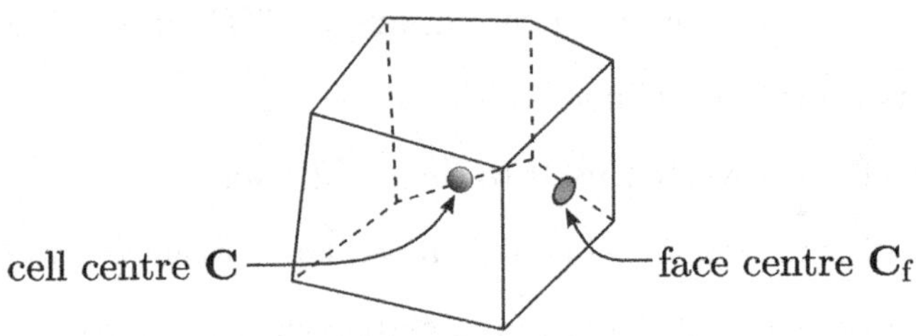


The finite volume method relates discrete values of fields, *e.g.* pressure, to cells and faces within the mesh. For many calculations data must to be assigned to point locations, in particular the cell centre (more specifically *centroid*)$\mathbf{C}$ and face centre $\mathbf{C}_{\mathrm{f}}$.

## Calculating mesh data

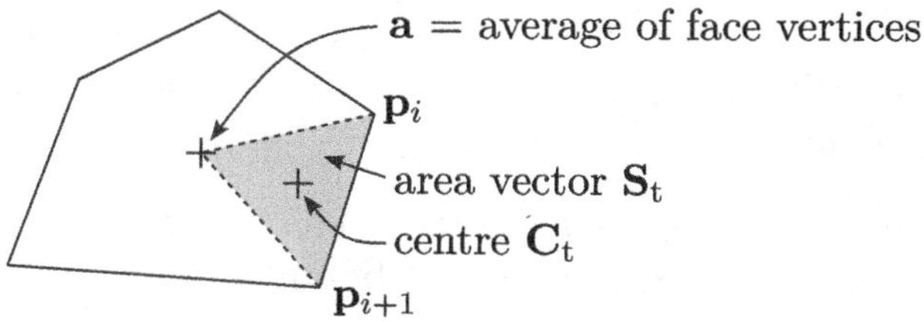


To calculate $\mathbf{C}_\mathrm{f}$, each polygonal face is decomposed into triangles using an apex point $\mathbf{a}$. The area vector $\mathbf{S}_\mathrm{t}$ and centre $\mathbf{C}_\mathrm{t}$ are then calculated for each triangle according to

$$\mathbf{S}_\mathrm{t} = (\mathbf{p}_{i+1} - \mathbf{p}_i) \times (\mathbf{a} - \mathbf{p}_i)/2$$
$$\mathbf{C}_\mathrm{t} = (\mathbf{p}_{i+1} + \mathbf{p}_i + \mathbf{a})/3,$$

where "$\times$" is the cross product, Eq. (2.70). The sum of area vectors gives $\mathbf{S}_\mathrm{f}$ and $\mathbf{C}_\mathrm{f}$ is the area weighted sum of triangle centres over $N$ triangles, calculated by

$$\mathbf{C}_\mathrm{f} = \frac{1}{|\mathbf{S}_\mathrm{f}|} \sum^{N} |\mathbf{S}_\mathrm{t}| \mathbf{C}_\mathrm{t}.$$

The cell volume $V$ can be calculated from $\mathbf{S}_\mathrm{f}$ and $\mathbf{C}_\mathrm{f}$ by Gauss's theorem, using $\mathbf{x}$ to describe position and noting $\nabla \bullet \mathbf{x} \equiv 3$. The surface integral ($\int_S$) becomes a sum over cell faces $\sum_\mathrm{f}$, replacing discrete values $\mathbf{S}_\mathrm{f}$ and $\mathbf{C}_\mathrm{f}$ for $d\mathbf{S}$ and $\mathbf{x}$ respectively as follows:

$$V = \int_V dV = \frac{1}{3} \int_V \nabla \bullet \mathbf{x} \, dV = \frac{1}{3} \int_S (d\mathbf{S} \bullet \mathbf{x}) \approx \frac{1}{3} \sum_\mathrm{f} \mathbf{S}_\mathrm{f} \bullet \mathbf{C}_\mathrm{f}.$$

The cell centre is calculated similarly, noting $\nabla |\mathbf{x}|^2 \equiv 2\mathbf{x}$, by

$$\mathbf{C} = \frac{1}{V} \int_V \mathbf{x} \, dV = \frac{1}{2V} \int_V \nabla |\mathbf{x}|^2 \, dV$$
$$= \frac{1}{2V} \int_S (d\mathbf{S} \, |\mathbf{x}|^2) \approx \frac{1}{2V} \sum_\mathrm{f} \mathbf{S}_\mathrm{f} \, |\mathbf{C}_\mathrm{f}|^2.$$

## 3.4 Equation discretisation

Equation discretisation converts partial differential equations for *continuous* fields, *e.g.* pressure $p$, into sets of linear equations for *discrete* fields.

The values of the principal fields, *e.g.* $p$, are associated with cells. A field is then represented by an array of values, $p_i$, for cell indices $i = 1, 2, 3, \ldots, N$. $N$ is the total number of cells.

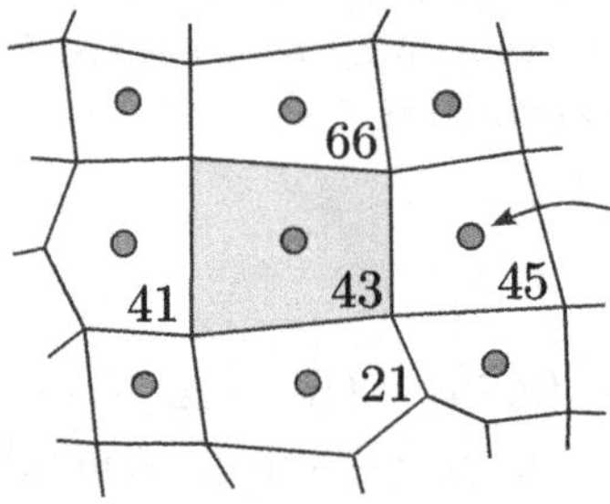


Equation discretisation creates a linear equation for each cell. For cell 43 above, the equation might have the following form:

$$a_{43,21}\, p_{21} + a_{43,41}\, p_{41} + \mathbf{a_{43,43}\, p_{43}} \\ + a_{43,45}\, p_{45} + a_{43,66}\, p_{66} = b_{43}.$$

where $a_{i,j}$ and $b_i$ are coefficients corresponding to cell indices $i$,$j$ (diagonal coefficient in **bold**). The set of linear equations for all cells can be written as a matrix equation of the form:

$$\begin{bmatrix} a_{1,1} & a_{1,2} & a_{1,3} & \cdots & a_{1,N} \\ a_{2,1} & a_{2,2} & a_{2,3} & \cdots & a_{2,N} \\ a_{3,1} & a_{3,2} & a_{3,3} & \cdots & a_{3,N} \\ \vdots & \vdots & \vdots & \ddots & \vdots \\ a_{N,1} & a_{N,2} & a_{N,3} & \cdots & a_{N,N} \end{bmatrix} \begin{bmatrix} p_1 \\ p_2 \\ p_3 \\ \vdots \\ p_N \end{bmatrix} = \begin{bmatrix} b_1 \\ b_2 \\ b_3 \\ \vdots \\ b_N \end{bmatrix} \tag{3.1}$$

The matrix contains of a set of coefficients $a_{i,j}$ where each row $i$ corresponds to the linear equation for the cell with index $i$. Each row of coefficients are non-zero only for the respective cell (diagonal $a_{i,i}$) and near-neighbours. All other coefficients are zero, making the matrix extremely *sparse*.

The matrix equation can be represented as $[\,\mathbf{A}\,][\,p\,] = [\,\mathbf{b}\,]$, where: $[\,\mathbf{A}\,]$ are the matrix coefficients; $[\,\mathbf{b}\,]$ are the source coefficients $b_i$; and, $[\,p\,]$ is the discretised pressure field. It may also be illustrated as follows, where '$\star$' indicates non-zero coefficients.

$$\begin{bmatrix} \star & \star & & \star \\ \star & \star & \star & \\ & \star & \star & \star \\ \star & & \star & \star \end{bmatrix} \begin{bmatrix} \\ p \\ \\ \end{bmatrix} = \begin{bmatrix} \star \\ \star \\ \star \\ \star \end{bmatrix}$$

## Segregated solution

A CFD simulation generally solves a set of physical equations, *e.g.* for mass, momentum and energy conservation. The finite volume method traditionally discretises each physical equation separately to form individual matrix equations for single solution variables, *e.g.* $[\,p\,]$, $[\,\mathbf{u}\,]$, $[\,T\,]$, rather than creating a single matrix equation that represents all the physical equations.

The *segregated* matrix equations are solved one variable at a time, *e.g.* solving for $[\,p\,]$, $[\,\mathbf{u}\,]$ and $[\,T\,]$ in separate steps. Where the solution variable is a vector or tensor, *e.g.* $\mathbf{u}$, it is *decoupled* into individual matrix equations for each component, *e.g.* $[\,u_x\,]$, $[\,u_y\,]$, $[\,u_z\,]$.

The matrix equations are solved in an *iterative* sequence, in which the equation for one variable, *e.g.* $[\,u_x\,]$, incorporates current values of other variables, *e.g.* $u_y$, $u_z$ and $p$, into the source vector, as shown below.

$$\begin{bmatrix} \star & \star & & \star \\ \star & \star & \star & \\ & \star & \star & \star \\ \star & & \star & \star \end{bmatrix} \begin{bmatrix} \\ u_x \\ \\ \end{bmatrix} = \begin{bmatrix} \star \\ \star \\ \star \\ \star \end{bmatrix} \longleftarrow \text{includes known } u_y,\, u_z,\, p$$

## 3.5 Matrix construction

The construction of each matrix equation involves building the matrix and source coefficients from the terms in the equation being solved, with further adjustments for boundary conditions.

The figure below illustrates the process of building a matrix equation for a field $\mathbf{\Psi}$ from an equation including advection, diffusion and source $S_{\mathbf{\Psi}}$.

$$\frac{\partial \mathbf{\Psi}}{\partial t} + \nabla \cdot (\mathbf{u}\mathbf{\Psi}) + \nabla \cdot (\Gamma \nabla \mathbf{\Psi}) = S_{\mathbf{\Psi}}$$

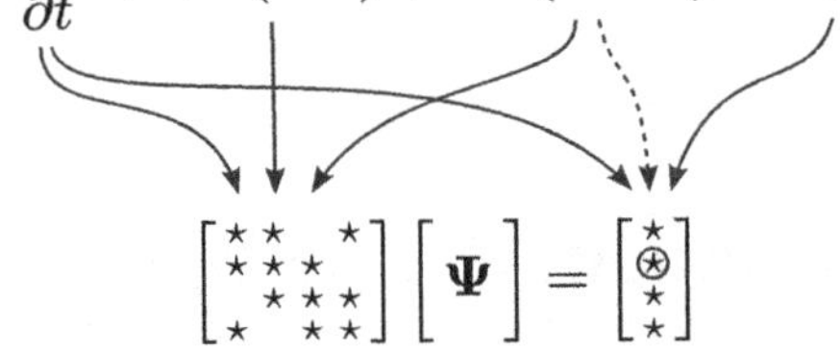


Coefficients for matrix $[\,\mathbf{A}\,]$ and source $[\,\mathbf{b}\,]$ are calculated for each individual term in the equation, *e.g.* $\partial \mathbf{\Psi}/\partial t$, $\nabla \cdot (\mathbf{u}\mathbf{\Psi})$, *etc.* using the discretisation methods described in this chapter.

The overall coefficients are calculated as the sum of coefficients for each term in the the equation. Most terms contribute to both the matrix and source coefficients, although this depends on the choice of discretisation scheme.

Finally, boundary conditions are incorporated into the equation through further adjustments to coefficients in $[\,\mathbf{A}\,]$ and $[\,\mathbf{b}\,]$ as shown below. The adjustments, principally from the advection and diffusion terms, are applied to coefficients corresponding to cells at the domain boundary.

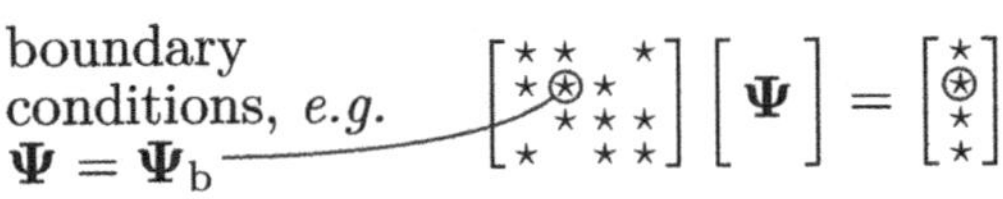

## Implicit and explicit

The equation for a field $\boldsymbol{\Psi}$ on the previous page is discretised to form the matrix equation $[\,\mathbf{A}\,][\,\boldsymbol{\Psi}\,] = [\mathbf{b}]$. The discretisation of a term is *implicit* when it contributes to coefficients in $[\,\mathbf{A}\,]$ by treating $\boldsymbol{\Psi}$ as the solved field $[\,\boldsymbol{\Psi}\,]$.

*Explicit* discretisation calculates coefficients in $[\mathbf{b}]$ only, using current values of fields. When solving an equation for $\boldsymbol{\Psi}$, derivatives without $\boldsymbol{\Psi}$ must be explicit. Terms with $\boldsymbol{\Psi}$ could be treated explicitly by using current values of $\boldsymbol{\Psi}$, but they generally are not, since explicit solutions are unstable beyond a limiting time step as described in Sec. 3.17. A notable exception to this are the terms discussed in Sec. 3.20.

The curl derivative, *e.g.* $\nabla \times \mathbf{u}$, includes terms in $u_y$ and $u_z$ in the decoupled matrix equation for $[\,u_x\,]$. These terms must be treated explicitly since they do not include $u_x$ itself. The situation for $[\,u_y\,]$ and $[\,u_z\,]$ is the same, so the curl derivative can be explicit only.

This leaves the following terms which are generally treated implicitly:[†]

- time derivative $\partial\boldsymbol{\Psi}/\partial t$;
- diffusion (Laplacian) $\nabla \cdot (\Gamma\nabla\boldsymbol{\Psi})$;
- advection $\nabla \cdot (\mathbf{u}\boldsymbol{\Psi})$;
- implicit linear function $c\boldsymbol{\Psi}$, where $c$ is a scalar.

This chapter details: the discretisation of these terms, which are generally treated implicitly; and, other terms such as divergence $\nabla \cdot \boldsymbol{\Psi}$ and gradient $\nabla\boldsymbol{\Psi}$ which can only be discretised explicitly within a segregated solution.

[†]also equivalent terms including density, *e.g.* $\partial(\rho\boldsymbol{\Psi})/\partial t$, $\nabla \cdot (\rho\mathbf{u}\boldsymbol{\Psi})$

## 3.6 Overview of discretisation

The choice of numerical method determines how coefficients for $[\mathbf{A}]$ and $[\mathbf{b}]$ are calculated and consequently the characteristics of the resulting matrix equation $[\mathbf{A}][\boldsymbol{\Psi}] = [\mathbf{b}]$.

The finite volume method described here is firmly rooted in the underlying concept of control volumes, described in Sec. 3.1. It uses integrals over a surface surrounding a volume applied to irregular polyhedral meshes, described in Sec. 3.2.

The discretisation is described in terms of differential operators, *e.g.* $\nabla \bullet (\Gamma \nabla \boldsymbol{\Psi})$ and $\nabla \bullet (\mathbf{u}\boldsymbol{\Psi})$, applied to a general field $\boldsymbol{\Psi}$.†

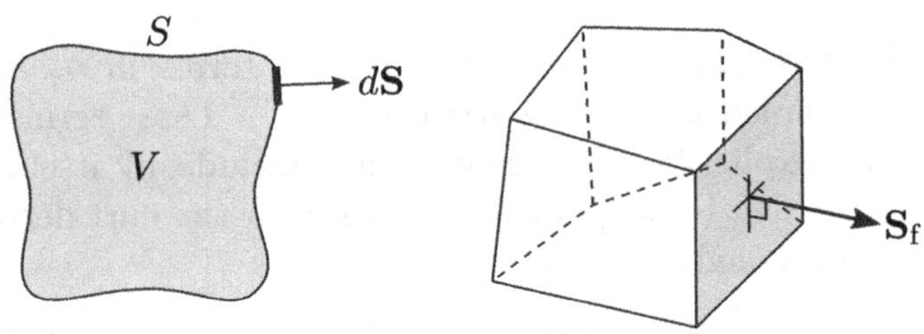


The main concept is that faces within a mesh form closed surfaces surrounding finite volumes, *e.g.* a single cell. Any surface integral that represents a derivative, *e.g.* $\nabla \bullet \boldsymbol{\Psi}$, is approximated by a summation over the faces 'f' that form the surface, *i.e.*

$$\int_S (d\mathbf{S} \bullet \boldsymbol{\Psi}) \rightarrow \sum_{\mathrm{f}} \mathbf{S}_{\mathrm{f}} \bullet \boldsymbol{\Psi}_{\mathrm{f}}.$$

The flux associated with $\boldsymbol{\Psi}$, *i.e.* $\mathbf{S}_{\mathrm{f}} \bullet \boldsymbol{\Psi}_{\mathrm{f}}$, must then be calculated. The value $\boldsymbol{\Psi}_{\mathrm{f}}$ is required at each *face*, which must be

†consistent with the idea of "Field Operation and Manipulation", which forms the acronym "FOAM" (later, OpenFOAM) used for the CFD software created by author Henry Weller.

calculated by some method of interpolation of values of $\boldsymbol{\Psi}$ from cells neighbouring the respective face.

## Intensive and extensive properties

In this chapter, derivatives and their discretisation are described *at a point*, *e.g.* $\nabla \bullet (\mathbf{u}\boldsymbol{\Psi})$, *e.g.* Eq. (3.8):

$$\nabla \bullet (\mathbf{u}\boldsymbol{\Psi}) \rightarrow \frac{1}{V} \sum_{\mathrm{f}} \phi_{\mathrm{f}} \boldsymbol{\Psi}_{\mathrm{f}}.$$

Calculating a derivative with this expression using the known field $\boldsymbol{\Psi}$ simply produces another field (with values defined at cell centres, with units of $\boldsymbol{\Psi}$/time).

The resulting field is *intensive*, meaning it is independent of the size of the system/geometry. Like other intensive fields, *e.g.* $\mathbf{u}$ and $\boldsymbol{\Psi}$ themselves, it can be used in further calculations, *e.g.* within another derivative or added/subtracted from other fields.

*Extensive* properties are dependent on the size of the system. For example, the volumetric flux $\phi_{\mathrm{f}} = \mathbf{S}_{\mathrm{f}} \bullet \mathbf{u}_{\mathrm{f}}$ described in Sec. 3.9 is dependent on the face areas $\mathbf{S}_{\mathrm{f}}$. Numerical operations involving extensive properties, *e.g.* addition, subtraction, or mapping to another location, generally produce meaningless data.

While *calculations* of derivatives yield intensive fields, *matrix equations are constructed in extensive form*, with coefficients and source vector scaled by cell volume $V$. In other words, in the discretisation example above, the multiplication by $(1/V)$ would be omitted.

There is no $(1/V)$ multiplier in the discretisation of terms which do not involve a surface integral, *e.g.* time derivative Eq. (3.21) and terms in Sec. 3.20. For those terms, the calculation of matrix coefficients and sources includes a multiplication by $V$.

## 3.7 Laplacian discretisation

Let us first describe the discretisation of the Laplacian term for diffusion, introduced in Sec. 2.14.

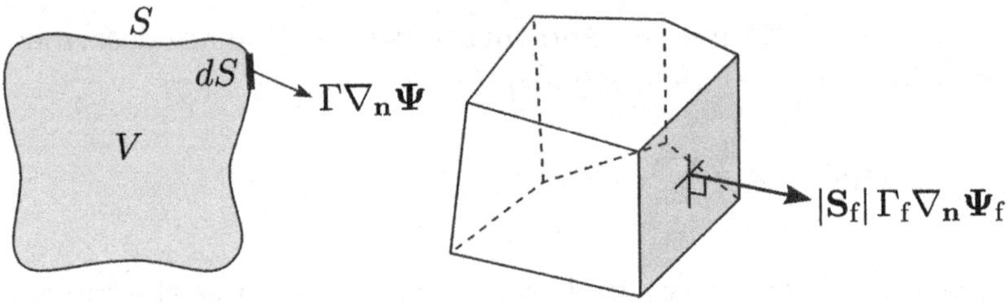


Following the finite volume principles described in Sec. 3.1, the discretisation approximates the surface integral by a summation over faces by

$$
\begin{aligned}
\nabla \bullet (\Gamma \nabla \boldsymbol{\Psi}) &= \lim_{\Delta V \to 0} \frac{1}{\Delta V} \int_S (dS\, \Gamma \nabla_{\mathbf{n}} \boldsymbol{\Psi}) \\
&\to \frac{1}{V} \sum_{\mathrm{f}} |\mathbf{S}_{\mathrm{f}}|\, \Gamma_{\mathrm{f}} \nabla_{\mathbf{n}} \boldsymbol{\Psi}_{\mathrm{f}}.
\end{aligned}
\tag{3.2}
$$

The $\sum_{\mathrm{f}}$ in Eq. (3.2) can be viewed as the summation over faces of a single cell. Applying the summation to all cells provides the contribution to coefficients $[\,\mathbf{A}\,]$ and $[\mathbf{b}]$ of a matrix equation.

The mesh data $V$ and $\mathbf{S}_{\mathrm{f}}$ are calculated according to Sec. 3.3 so the remaining properties to be determined are:

- the diffusivity at faces $\Gamma_{\mathrm{f}}$;
- the surface normal gradient at faces $\nabla_{\mathbf{n}} \boldsymbol{\Psi}_{\mathrm{f}}$.

Fields $\Gamma$ and $\boldsymbol{\Psi}$ are associated with cells, so numerical schemes are required to evaluate properties at faces. We will first describe interpolation for $\Gamma_{\mathrm{f}}$, for which the linear scheme is generally used. The surface normal gradient $\nabla_{\mathbf{n}} \boldsymbol{\Psi}_{\mathrm{f}}$ is discussed in Sec. 3.8.

## Interpolation from cells to faces

Mapping data between different locations is a common practice in numerics. Since the finite volume method is concerned with fluxes at faces, the principal mapping procedure is *interpolation* from cells to faces. Since other interpolations are much less common, it can be assumed that the term "interpolation" means from cell to face unless stated otherwise.

For irregular polyhedral meshes, interpolation is generalised by defining a *weights* field $w$ for each face according to

$$\boldsymbol{\Psi}_{\mathrm{f}} = w\,\boldsymbol{\Psi}_{\mathrm{P}} + (1 - w)\,\boldsymbol{\Psi}_{\mathrm{N}}, \tag{3.3}$$

where $\boldsymbol{\Psi}_{\mathrm{f}}$ is the interpolated face field. The subscripts P and N indicate values at owner and neighbour cells, respectively.

## Linear interpolation

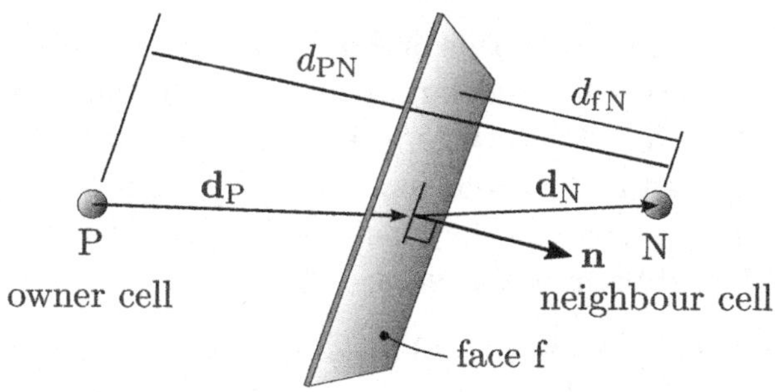


The **linear** interpolation scheme sets $\boldsymbol{\Psi}_{\mathrm{f}}$ according to a linear variation between cells values $\boldsymbol{\Psi}_{\mathrm{P}}$ and $\boldsymbol{\Psi}_{\mathrm{N}}$. The weights can then be calculated based on distances from the face centre to adjacent cell centres, in the direction normal to the face, by

$$w = \frac{d_{\mathrm{fN}}}{d_{\mathrm{PN}}} = \frac{\mathbf{n} \bullet \mathbf{d}_{\mathrm{N}}}{\mathbf{n} \bullet (\mathbf{d}_{\mathrm{P}} + \mathbf{d}_{\mathrm{N}})}. \tag{3.4}$$

## 3.8 Surface normal gradient

The surface normal gradient $\nabla_{\mathbf{n}}\boldsymbol{\Psi}_{\mathrm{f}}$ is a part of the Laplacian discretisation Eq. (3.2), illustrated in the figure below.

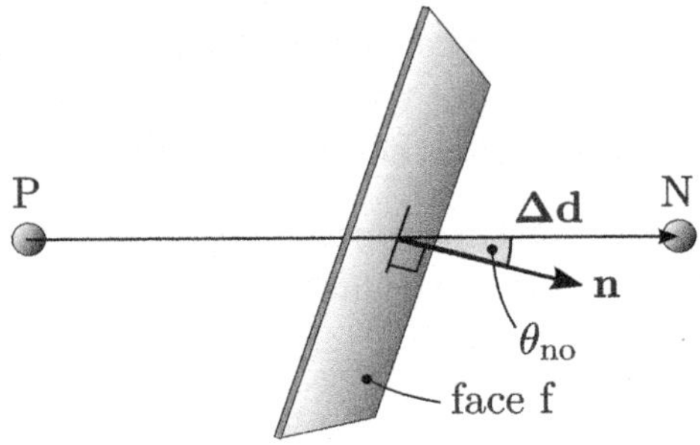


The discretisation of $\nabla_{\mathbf{n}}\boldsymbol{\Psi}_{\mathrm{f}}$ is built upon a finite difference of cell values on each side of the face according to

$$\nabla_{\mathbf{n}}\boldsymbol{\Psi}_{\mathrm{f}} = C_{\Delta}(\boldsymbol{\Psi}_{\mathrm{N}} - \boldsymbol{\Psi}_{\mathrm{P}}), \tag{3.5}$$

where $C_{\Delta} = 1/|\boldsymbol{\Delta}\mathbf{d}|$. When this **orthogonal** scheme is applied to Eq. (3.2) to discretise a Laplacian, it forms coefficients $[\,\mathbf{A}\,]$ of a matrix equation $[\,\mathbf{A}\,][\,\boldsymbol{\Psi}\,] = [\mathbf{b}]$ since it references cell values of the field $\boldsymbol{\Psi}$. For cell $i$, the coefficient for each neighbour cell (N) is $a_{i,\mathrm{N}} = |\mathbf{S}_{\mathrm{f}}|\,\Gamma_{\mathrm{f}}\,C_{\Delta}$ and the diagonal coefficient is the negative of the sum of neighbour coefficients: $-\sum_{\mathrm{N}} a_{i,\mathrm{N}}$.

Discretisation of $\nabla_{\mathbf{n}}\boldsymbol{\Psi}_{\mathrm{f}}$ by Eq. (3.5) is most accurate when the face is orthogonal to $\boldsymbol{\Delta}\mathbf{d}$, *i.e.* the angle $\theta_{\mathrm{no}}$ between $\mathbf{n}$ and $\boldsymbol{\Delta}\mathbf{d}$ is zero. However, if the face is *non-orthogonal*, the error associated with Eq. (3.5) increases with $\theta_{\mathrm{no}}$.

### Non-orthogonal correction

A more accurate discretisation of $\nabla_{\mathbf{n}}\boldsymbol{\Psi}_{\mathrm{f}}$ at a non-orthogonal face is formed of the vector sum of the orthogonal scheme $\nabla_{\mathbf{n}}\boldsymbol{\Psi}_{\mathrm{f}}^{\mathrm{orth}}$

and an *explicit* correction $\nabla_{\mathbf{n}}\boldsymbol{\Psi}_{\mathrm{f}}^{\mathrm{corr}}$. The latter is calculated from the full gradient $\nabla\boldsymbol{\Psi}$ in adjacent cells (described in Sec. 3.15), interpolated to the face $(\nabla\boldsymbol{\Psi})_{\mathrm{f}}$.

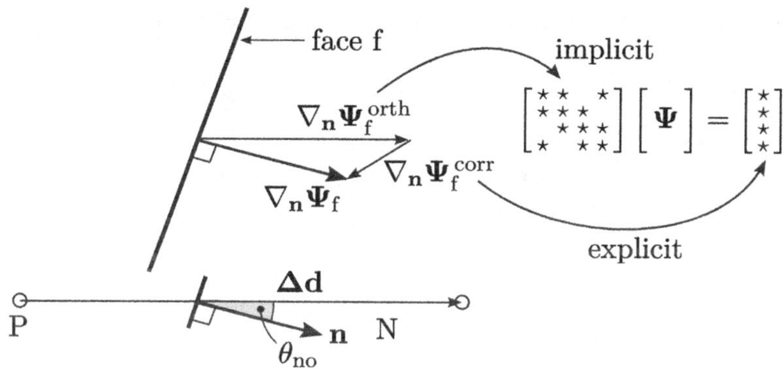


The correction $\nabla_{\mathbf{n}}\boldsymbol{\Psi}_{\mathrm{f}}^{\mathrm{corr}}$ is explicit, *i.e.* calculated using known values of $\boldsymbol{\Psi}$, so may need updating within an iterative sequence to maintain accuracy, as discussed in Sec. 5.20. To ensure that the iterative sequence converges, the implicit contribution is elevated by replacing $C_{\Delta}$ in the orthogonal scheme with

$$C_{\Delta}^{\mathrm{corr}} = \frac{1}{\mathbf{n}\bullet\boldsymbol{\Delta}\mathbf{d}} = \frac{1}{|\boldsymbol{\Delta}\mathbf{d}|\,\cos\theta_{\mathrm{no}}}. \tag{3.6}$$

The **corrected** $\nabla_{\mathbf{n}}\boldsymbol{\Psi}_{\mathrm{f}}$ scheme combines the implicit and explicit parts by

$$\nabla_{\mathbf{n}}\boldsymbol{\Psi}_{\mathrm{f}} = \underbrace{C_{\Delta}^{\mathrm{corr}}(\boldsymbol{\Psi}_{\mathrm{N}} - \boldsymbol{\Psi}_{\mathrm{P}})}_{\text{orthogonal, implicit}} + \underbrace{(\mathbf{n} - C_{\Delta}^{\mathrm{corr}}\boldsymbol{\Delta}\mathbf{d})\bullet(\nabla\boldsymbol{\Psi})_{\mathrm{f}}}_{\text{correction, explicit}}. \tag{3.7}$$

The corrected scheme is generally stable for $\theta_{\mathrm{no}} \lesssim 75°$. For $\theta_{\mathrm{no}} > 75°$, stability can be maintained at the expense of accuracy by *limiting* the magnitude of the correction $\nabla_{\mathbf{n}}\boldsymbol{\Psi}_{\mathrm{f}}^{\mathrm{corr}}$ below some fraction of the magnitude of the orthogonal $\nabla_{\mathbf{n}}\boldsymbol{\Psi}_{\mathrm{f}}^{\mathrm{orth}}$ part.

## 3.9 Advection discretisation

In Sec. 2.8, we described advection terms as those of the form $\nabla \cdot (\mathbf{u}\boldsymbol{\Psi})$ or $\nabla \cdot (\rho\mathbf{u}\boldsymbol{\Psi})$. The inclusion of velocity $\mathbf{u}$ within the divergence gives the term particular characteristics that require special treatment in discretisation.

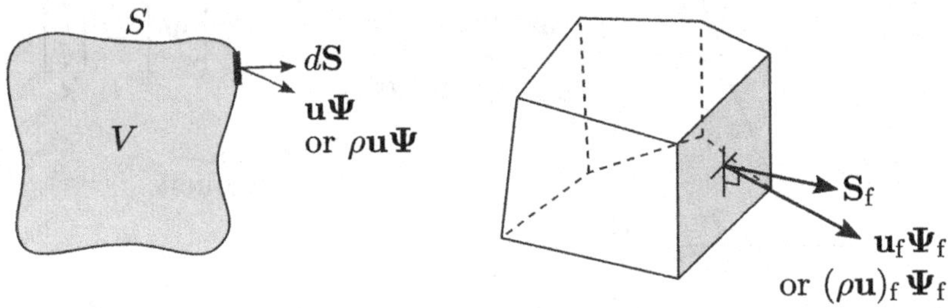


Following the finite volume principles outlined in Sec. 3.1, the discretisation approximates the surface integral by a summation over faces by

$$\begin{aligned}\nabla \cdot (\mathbf{u}\boldsymbol{\Psi}) &= \lim_{\Delta V \to 0} \frac{1}{\Delta V} \int_S (d\mathbf{S} \cdot \mathbf{u}\boldsymbol{\Psi}) \\ &\to \frac{1}{V} \sum_{\mathrm{f}} \mathbf{S}_{\mathrm{f}} \cdot \mathbf{u}_{\mathrm{f}} \boldsymbol{\Psi}_{\mathrm{f}} \to \frac{1}{V} \sum_{\mathrm{f}} \phi_{\mathrm{f}} \boldsymbol{\Psi}_{\mathrm{f}}.\end{aligned} \tag{3.8}$$

The discretisation requires calculation of the volumetric flux $\phi_{\mathrm{f}} = \mathbf{S}_{\mathrm{f}} \cdot \mathbf{u}_{\mathrm{f}}$ (see Sec. 2.3). In the case of $\rho$ included in the advection term $\nabla \cdot (\rho\mathbf{u}\boldsymbol{\Psi})$, $\phi_{\mathrm{f}}$ is the mass flux $\mathbf{S}_{\mathrm{f}} \cdot (\rho\mathbf{u})_{\mathrm{f}}$.

In the flux calculation, the interpolation of $\mathbf{u}$ at cell centres to $\mathbf{u}_{\mathrm{f}}$ at faces uses the linear scheme of Sec. 3.7. Similarly, the linear scheme is used for the interpolation $\rho\mathbf{u} \to (\rho\mathbf{u})_{\mathrm{f}}$.

The critical issue — one of the most important in CFD numerics — is how to express our advected property $\boldsymbol{\Psi}_{\mathrm{f}}$ at a face in terms of values $\boldsymbol{\Psi}$ in neighbouring cells.

## Advection scheme introduction

The advected property $\boldsymbol{\Psi}$ is transported in the direction of flow velocity $\mathbf{u}$. Interpolation to the face of $\boldsymbol{\Psi}_\mathrm{f}$ usually involves the flow direction. In the graphic below a face f is positioned between two cells. Based on the flow direction, the cells are labelled upwind 'U' and downwind 'D'.

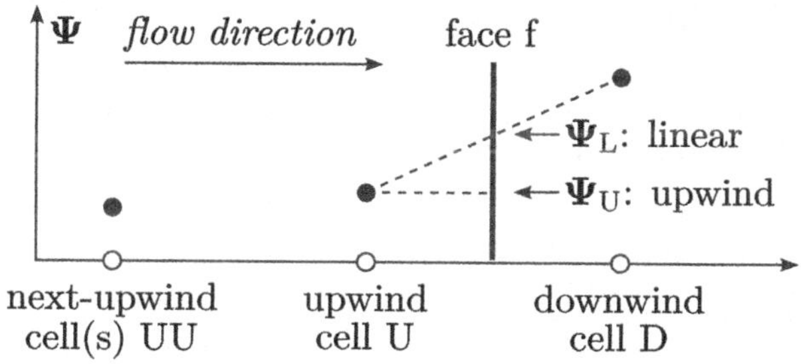


The **linear** interpolation scheme,[†] described in Sec. 3.7, does not use the flow direction but expresses $\boldsymbol{\Psi}_\mathrm{f}$ in terms of adjacent cells. At first sight, this choice of scheme is logical when considering *accuracy*. However, for advection, the linear scheme tends to generate unbounded solutions which are unstable.

The **upwind** scheme simply represents $\boldsymbol{\Psi}_\mathrm{f}$ by the value in the upwind cell $\boldsymbol{\Psi}_\mathrm{U}$. It makes sense from a physical perspective since particles of fluid in the upwind cell are destined to travel to the face, transporting property $\boldsymbol{\Psi}_\mathrm{U}$ with them.[‡]

While the linear scheme is generally unbounded, the upwind scheme exhibits poor accuracy. In following section we explore the behaviour of upwind before looking at schemes that offer greater accuracy while attempting to maintain boundedness.

[†]also known as the *central difference* scheme, particularly in the context of advection.

[‡]or, to quote Brian Spalding: "the wind from a pigsty always stinks".

## 3.10 Upwind scheme

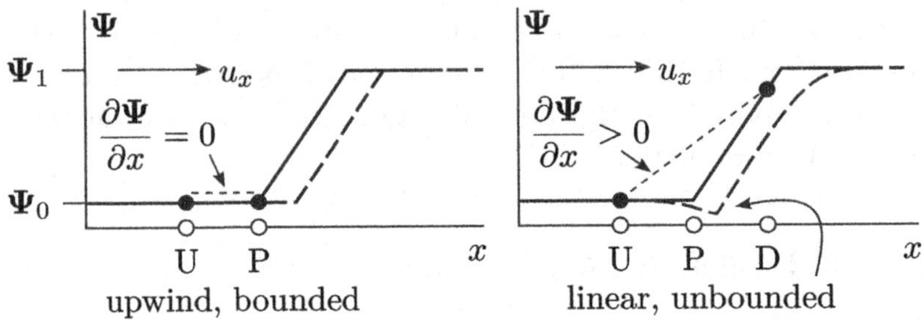


The upwind scheme represents the face value $\mathbf{\Psi}_\mathrm{f}$ by the value $\mathbf{\Psi}_\mathrm{U}$ in the cell upwind of the face. The advantage of upwind is that it can *guarantee* boundedness of a field $\mathbf{\Psi}$. We can demonstrate this point by revisiting the 1D Eq. (2.32) in Sec. 2.9:

$$\frac{\partial \mathbf{\Psi}}{\partial t} + u_x \frac{\partial \mathbf{\Psi}}{\partial x} = 0.$$

In the graphic above, we track the translation of a profile of $\mathbf{\Psi}$ by equating changes in $\mathbf{\Psi}$ in time to the local gradient $\partial\mathbf{\Psi}/\partial x$. If we apply upwind to calculate the change at point P, the gradient $\partial\mathbf{\Psi}/\partial x = 0$ and no change in $\mathbf{\Psi}$ is correctly calculated (left).

However, linear differencing between upwind and downwind values results in $\partial\mathbf{\Psi}/\partial x > 0$, so predicts a decrease in the value at P (right). The solution produces a solution with $\mathbf{\Psi} < \mathbf{\Psi}_0$, so is unbounded.

Boundedness of the conservative form of advection $\nabla \cdot (\mathbf{u}\mathbf{\Psi})$ is only guaranteed when $\nabla \cdot \mathbf{u} = 0$, as discussed in Sec. 2.9. In 1D, the conservative form moves $u_x$ inside the derivative $\partial(u_x\mathbf{\Psi})/\partial x$. That gradient is only zero with upwind when $u_x$ is uniform, *i.e.* the 1D equivalent to $\nabla \cdot \mathbf{u} = 0$.

## Diffusion of upwind

The upwind scheme is highly *diffusive* which can result in poor accuracy. Its diffusive nature can be explained by considering the following Taylor's series expansion:

$$\boldsymbol{\Psi}(x + \Delta x) = \boldsymbol{\Psi}(x) + \frac{\partial \boldsymbol{\Psi}}{\partial x}\Delta x + \frac{\partial^2 \boldsymbol{\Psi}}{\partial x^2}\frac{\Delta x^2}{2} + \ldots \qquad (3.9)$$

In our 1D example, the upwind scheme calculates $\partial \boldsymbol{\Psi}/\partial x$ using $\boldsymbol{\Psi}_{\mathrm{U}}$ and $\boldsymbol{\Psi}_{\mathrm{P}}$ at locations U and P, separated by distance $\Delta x$. Relating the upwind calculation to Eq. (3.9) gives

$$\frac{\boldsymbol{\Psi}_{\mathrm{P}} - \boldsymbol{\Psi}_{\mathrm{U}}}{\Delta x} = \frac{\partial \boldsymbol{\Psi}}{\partial x} + \frac{\partial^2 \boldsymbol{\Psi}}{\partial x^2}\frac{\Delta x}{2} + \ldots \qquad (3.10)$$

In other words, the upwind discretisation represents $\partial \boldsymbol{\Psi}/\partial x$ but also the second derivative $\partial^2 \boldsymbol{\Psi}/\partial x^2$ (and higher derivatives). $\partial^2 \boldsymbol{\Psi}/\partial x^2$ is equivalent to a Laplacian, described in Sec. 2.14, which diffuses $\boldsymbol{\Psi}$ with a diffusivity proportional to $\Delta x$.

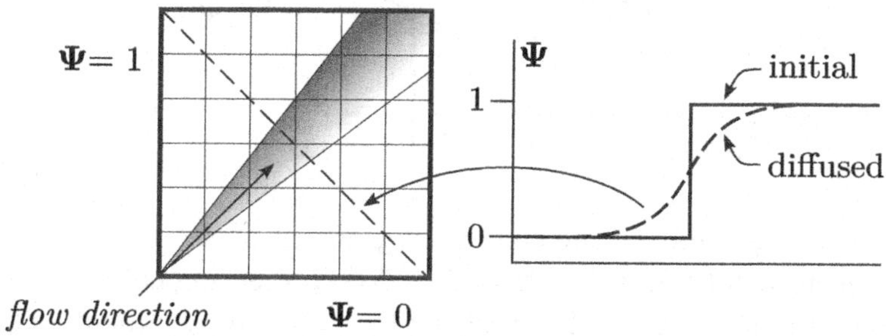


The upwind scheme is particularly diffusive when the flow direction is not aligned with the cells of a mesh. In the 2D box of cells above, $\boldsymbol{\Psi}$ is advected at a 45° angle, beginning with an abrupt step change from $\boldsymbol{\Psi} = 1$ and $\boldsymbol{\Psi} = 0$ between the left and lower boundaries. The step rapidly diffuses along the direction of travel as shown in graph (right) and shaded area (left).

## 3.11 Limited advection schemes

Alternative schemes for advection attempt to overcome problems with boundedness and accuracy of the linear and upwind schemes respectively. Many schemes apply a *limiter* $\beta$ between $\mathbf{\Psi}_{\mathrm{U}}$ from upwind and $\mathbf{\Psi}_{\mathrm{L}}$ from the linear scheme Eq. (3.4) according to

$$\mathbf{\Psi}_{\mathrm{f}} = (1 - \beta)\mathbf{\Psi}_{\mathrm{U}} + \beta\mathbf{\Psi}_{\mathrm{L}}. \tag{3.11}$$

When $\beta = 0$, the scheme reduces to upwind and it becomes linear interpolation when $\beta = 1$. For a uniform mesh ($w = 0.5$), $\beta = 2$ represents interpolation using the downwind cell value.

Limited schemes attempt to optimise $\beta$ at each face, based on the local $\mathbf{\Psi}$, to maximise accuracy whilst maintaining boundedness.

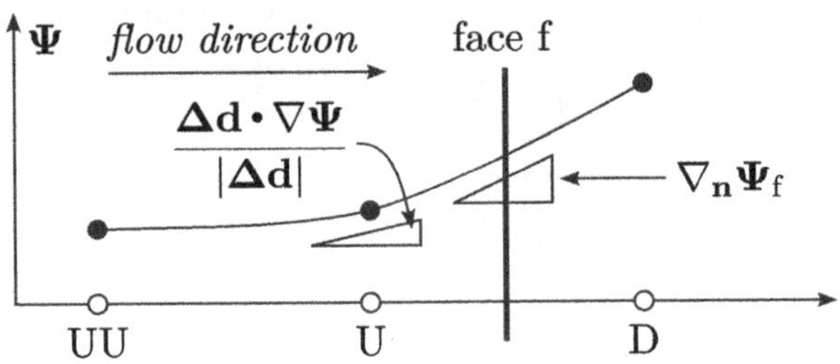


Many schemes analyse the change in gradient of $\mathbf{\Psi}_{\mathrm{f}}$ between the face and upwind cell in the direction $\mathbf{\Delta d}$ connecting cell centres. They define a function of a ratio $r$ of consecutive gradients as:

$$r = \max\left[2\frac{\mathbf{\Delta d} \bullet \nabla\mathbf{\Psi}}{|\mathbf{\Delta d}|\,\nabla_{\mathbf{n}}\mathbf{\Psi}_{\mathrm{f}}} - 1, 0\right] \quad \text{for scalar } \mathbf{\Psi}. \tag{3.12}$$

There are numerous published schemes that define the limiter $\beta(r)$ as a function of the gradient ratio $r$. Those schemes that are most useful are described in the following sections.

## Total variation diminishing schemes

Many useful schemes fall into a class known as Total Variation Diminishing (TVD).[†] The TVD idea is that if the total variation of field $\boldsymbol{\Psi}$ does not increase in time, "overshoots" and oscillations associated with unboundedness will not occur.

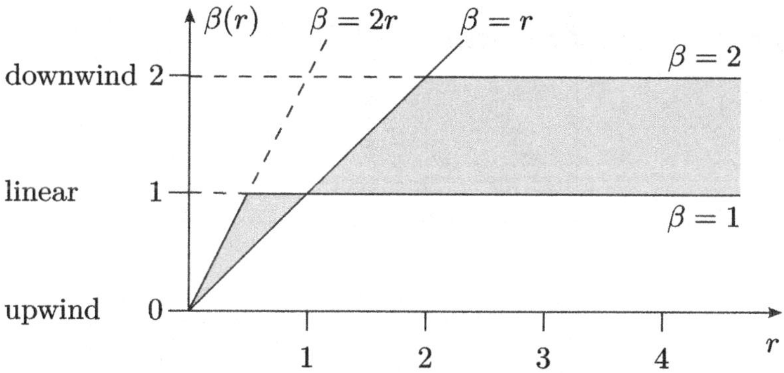


To qualify as TVD, the limiter function $\beta(r)$ must fall within the shaded area in a Sweby diagram (above).[‡] The TVD concept is a 1D analysis. For 3D CFD on irregular polyhedral meshes, oscillations are more likely to occur with TVD schemes whose $\beta(r)$ functions tend significantly to downwind, *i.e.* towards the upper part of the shaded area near $\beta = 2$.

A further property of a limited scheme is *symmetry*. A scheme is symmetric when the condition $\beta(r)/r = \beta(1/r)$ is satisfied. When this occurs, the scheme applies the same limiter to the gradient of $\boldsymbol{\Psi}$, irrespective of the *sign* of its gradient. As a consequence, a property $\boldsymbol{\Psi}$, initialised with a symmetric profile, *e.g.* a bell curve, will retain its symmetry under advection.

[†] Ami Harten, *High resolution schemes for hyperbolic conservation laws*, 1983.

[‡] Peter Sweby, *High resolution schemes using flux limiters for hyperbolic conservation laws*, 1984.

## 3.12 Useful TVD schemes

For irregular 3D polyhedral meshes, only TVD schemes whose limiter function $\beta(r)$ avoids excessive downwind interpolation, are sufficiently robust to be practically useful.

### Limited linear and minmod

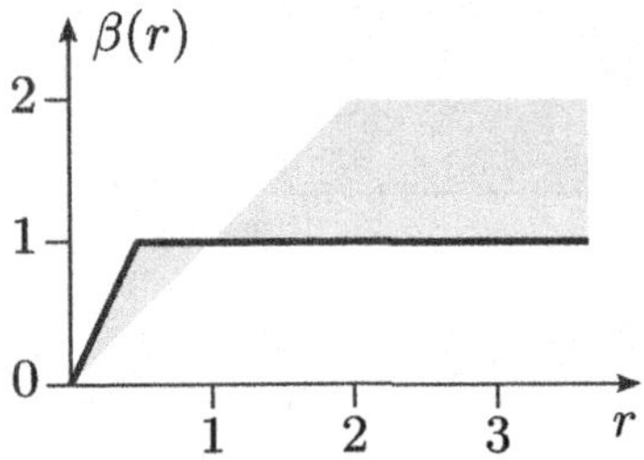

limited linear

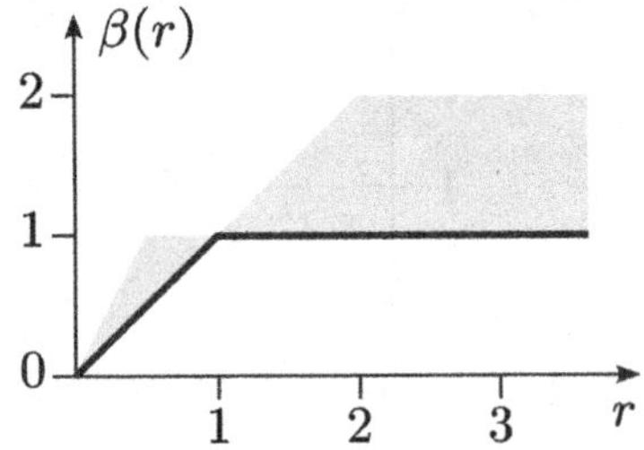

minmod

The **limited linear** scheme — as the name suggests — maintains the linear scheme as much as possible to be TVD, before limiting to upwind to improve boundedness, according to

$$\beta = \min(2r, 1)\,. \tag{3.13}$$

Limited linear is a robust scheme that exhibits low diffusion, but has a disadvantage that it is not symmetric, according to the criterion described in Sec. 3.11.

Like limited linear, the **minmod** scheme† does not introduce any downwind interpolation. It is symmetric but is more diffusive by limiting more readily towards upwind, according to

$$\beta = \min(r, 1)\,. \tag{3.14}$$

†Philip Roe, *Characteristic-based schemes for the Euler equations*, 1986.

## van Leer and van Albada

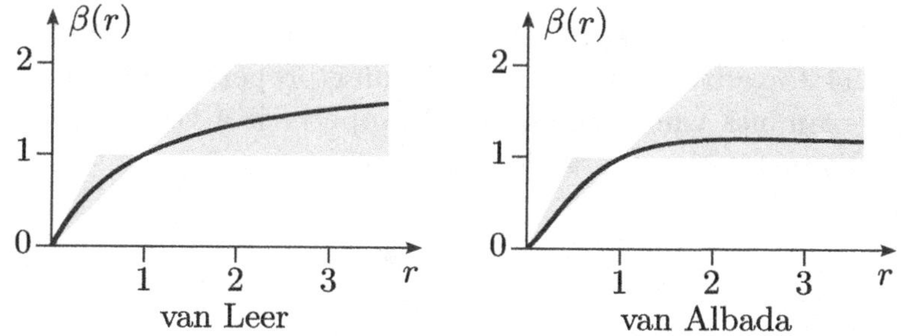


There are two notable TVD schemes that: exhibit low diffusion; maintain symmetry through some downwind interpolation; and, have limiter functions which are continuous in gradient which avoids problems due to sudden changes in limiter values.

The **van Leer** scheme† is the original limited advection scheme described by the limiter function

$$\beta = \frac{r + |r|}{1 + |r|}. \tag{3.15}$$

Since $\beta \to 2$ as $r \to \infty$, the van Leer scheme uses relatively strong downwind interpolation.

The **van Albada** scheme‡ is described by the limiter function

$$\beta = \frac{r^2 + r}{r^2 + 1}. \tag{3.16}$$

It uses much less downwind interpolation than the van Leer scheme since $\beta \to 1$ as $r \to \infty$. From the diagram, it appears like a smooth-functioned, less diffuse version of minmod.

---

†Bram van Leer, *Towards the ultimate conservative difference scheme II. Monotonicity and conservation combined in a second order scheme*, 1974.

‡Gale van Albada, *et al. A comparative study of computational methods in cosmic gas dynamics*, 1982.

## 3.13 Limiting multiple components

The calculation and application of a limiter is introduced in Sec. 3.11 for advection of a single scalar property. The figure below examines when the advected property is a vector, *e.g.* $\mathbf{u}$ in the momentum conservation Eq. (2.47).

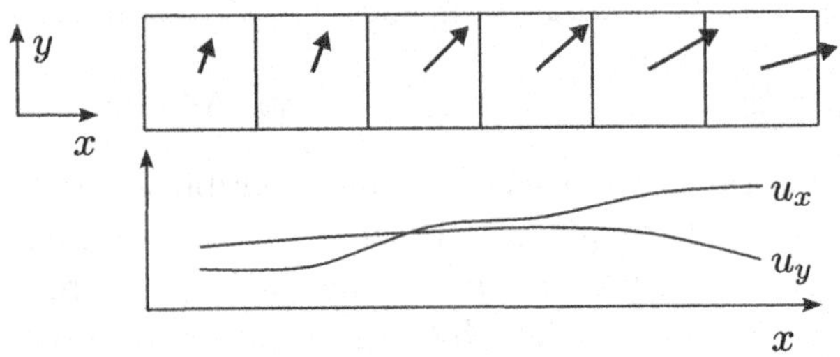


It shows a 2D cut along the $x$-$y$ plane through a selection of cells in a regular mesh with the velocity vector $\mathbf{u}$ displayed for each cell. When $\mathbf{u}$ is interpolated to faces along the $x$-direction using a limited scheme, the simplest approach is to calculate a limiter for each vector component, *e.g.* $u_x$, and interpolate the component with that limiter.

In the example above, the profiles of $u_x$ and $u_y$ in the $x$-direction are quite different, so the calculated limiters will be different for $u_x$ and $u_y$ components at each face. The limiting will **not** be *invariant* under a rotation of the co-ordinate axes, leading to a different solution depending on the initial orientation of the geometry with respect to the axes.

The limiting can be invariant using a single limiter calculated from the magnitude $|\mathbf{u}|$, which is applied to all components of $\mathbf{u}$. The strength of the limiting corresponds to an average across the components, which is usually insufficient for the component which requires strongest limiting. This can cause instability.

Instead, the **'V' scheme** calculates the limiter based on the 'worst-case' direction, *i.e.* the direction of steepest gradient in $\mathbf{\Psi}$ at the cell face. It uses the following expression for $r$ for a vector $\mathbf{\Psi}$, replacing Eq. (3.12) for a scalar:

$$r = 2\frac{\nabla_{\mathrm{n}}\mathbf{\Psi}_{\mathrm{f}} \bullet \mathbf{\Delta d} \bullet \nabla\mathbf{\Psi}}{|\mathbf{\Delta d}|\,\nabla_{\mathrm{n}}\mathbf{\Psi}_{\mathrm{f}} \bullet \nabla_{\mathrm{n}}\mathbf{\Psi}_{\mathrm{f}}} - 1, \quad \text{for vector } \mathbf{\Psi}. \tag{3.17}$$

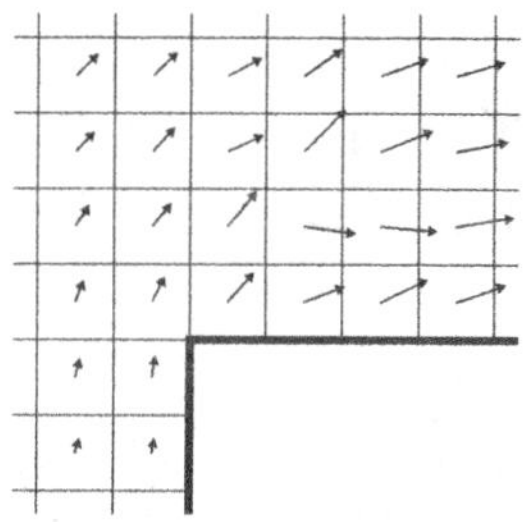
without 'V'-scheme

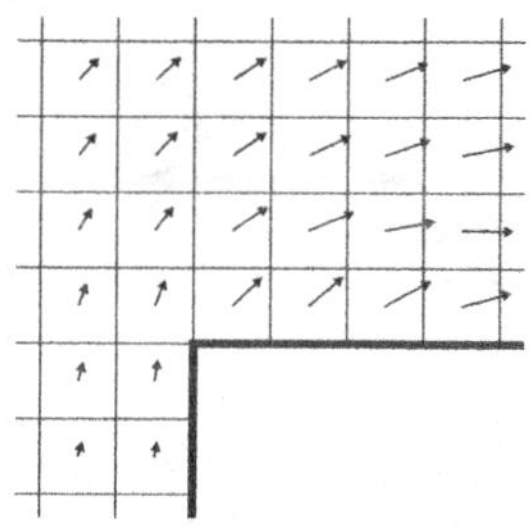
with 'V'-scheme

While the V scheme ensures invariance, it also provides greater stability than component limiting. It can remove oscillations in solutions, *e.g.* in the example above of supersonic flow over a step showing the effect in velocity in the cells adjacent to the step corner.

## Multivariate limiting

Multivariate limiting applies the same limiter for advection discretisation across a set of 2 or more equations. It works by calculating the limiter $\beta$ for each solution variable in the equation set and applying the lowest $\beta$ to all equations.

It can be used in order to maintain consistency in the transport of individual fluid species, such as $N_2$, $O_2$, *e.g.* in the propagation of laminar flame (which is beyond the scope of this book).

## 3.14 Linear upwind scheme

**Linear upwind** is another significant scheme for advection. It is used particularly for advection of momentum, *i.e.* $\nabla \bullet (\mathbf{uu})$ or $\nabla \bullet (\rho \mathbf{uu})$ but can be effective for advection of other variables.

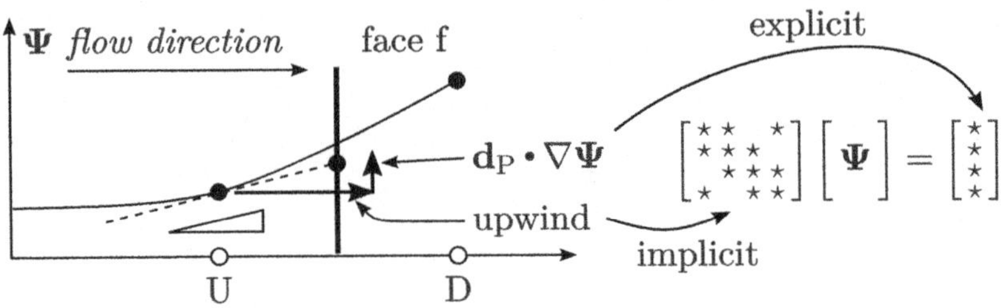


The scheme reduces the diffusive nature of upwind by including additional upwind cell values indirectly from the gradient $\nabla \mathbf{\Psi}$ in the upwind cell.

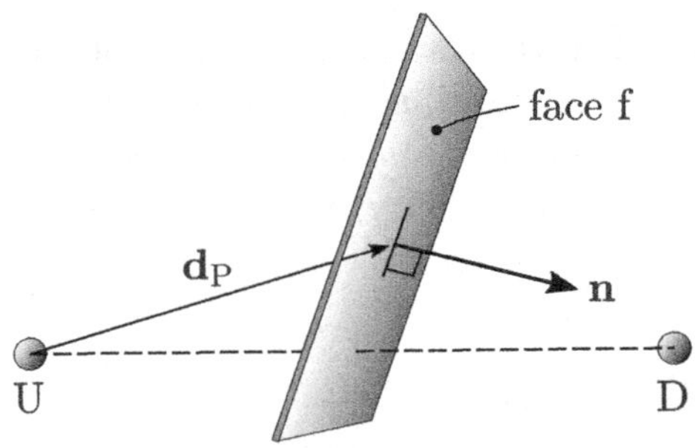


Linear upwind describes the face value as an extrapolation of the upwind cell value to the face using the upwind cell gradient $\nabla \mathbf{u}$ and a vector $\mathbf{d}_\mathrm{P}$ from the cell centre to face centre. It first provides a contribution to the coefficients $[\mathbf{A}]$ of a matrix equation $[\mathbf{A}][\mathbf{\Psi}] = [\mathbf{b}]$ by representing face values $\mathbf{\Psi}_\mathrm{f}$ by the upwind value $\mathbf{\Psi}_\mathrm{U}$. The extrapolation is then introduced through an additional *explicit* contribution $\mathbf{d}_\mathrm{P} \bullet \nabla \mathbf{u}$ to $[\mathbf{b}]$.

## Skewness

Interpolation of values between cell centres is along a line joining the cell centres. Any interpolated value at a face relates to the point on the face intersected by that line.

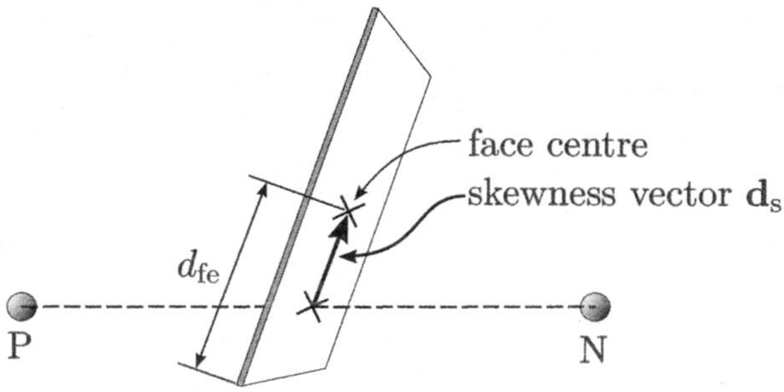


*Skewness* is the distance between the intersection point and the face centre. It can be represented by a vector $\mathbf{d}_\mathrm{s}$ or as the ratio $|\mathbf{d}_\mathrm{s}|/d_\mathrm{fe}$, where $d_\mathrm{fe}$ is the distance from the face centre to edge in the direction of $\mathbf{d}_\mathrm{s}$.

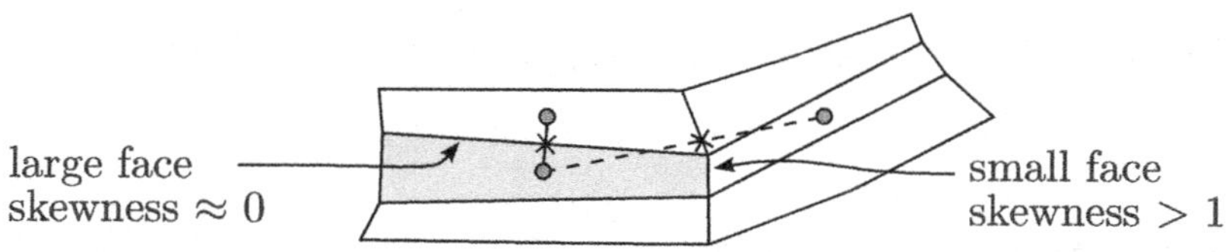


An interpolated value represents an average across a face, but as skewness increases, this representation becomes less accurate. High skewness (*e.g.* $> 1$) does not immediately equate to poor accuracy, however, since interpolated values $\mathbf{\Psi}_\mathrm{f}$ are multiplied by face areas within a surface integral, *e.g.* $\sum_\mathrm{f} |\mathbf{S}_\mathrm{f}| \mathbf{\Psi}_\mathrm{f}$ — and high skewness occurs at small faces.

Advection schemes do not generally include a correction for high skewness to improve accuracy. However, the linear upwind scheme naturally includes skewness correction since the explicit contribution is in the direction $\mathbf{d}_\mathrm{P}$ towards the face centre.

## 3.15 Gradient discretisation

The discretisation of a gradient $\nabla\boldsymbol{\Psi}$ is exclusively an explicit calculation using current values of $\boldsymbol{\Psi}$. The conservative form of gradient calculation is based on a surface integral.

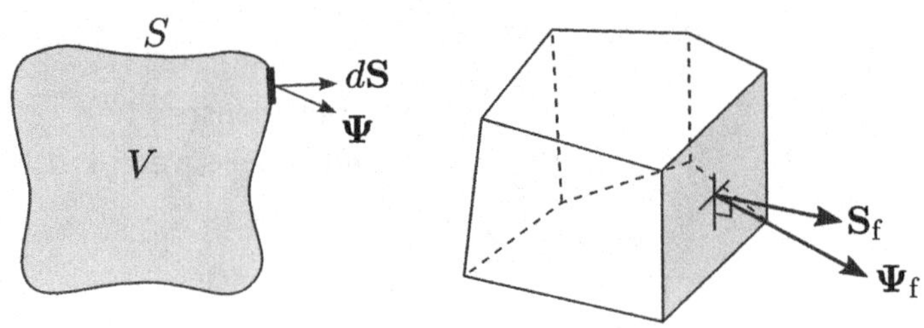


From the gradient definition in Sec. 2.23, the discretisation is

$$\nabla\boldsymbol{\Psi} = \lim_{\Delta V \to 0} \frac{1}{\Delta V} \int_S (d\mathbf{S}\,\boldsymbol{\Psi}) \to \frac{1}{V} \sum_{\mathrm{f}} \mathbf{S}_{\mathrm{f}} \boldsymbol{\Psi}_{\mathrm{f}}. \tag{3.18}$$

The face value $\boldsymbol{\Psi}_{\mathrm{f}}$ is generally interpolated from cell values using the linear scheme.

### Point linear interpolation

While skewness is generally not a concern for advection discretisation, it deserves greater attention in gradient calculation. For "bad" meshes, *e.g.* containing elongated tetrahedral cells, the point linear scheme is often adopted to reduce skewness error.

**Point linear** interpolation uses: the value $\boldsymbol{\Psi}_{\mathrm{f}}$, calculated using linear interpolation, which corresponds to the "face point" at the intersection of the line connecting cell centres and the face; and, values $\boldsymbol{\Psi}_{\mathrm{V}}$ at each vertex, interpolated from adjacent cells using inverse-distance weighting.

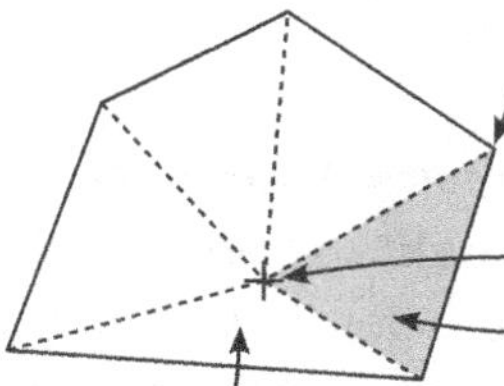


The scheme breaks the polygonal face into triangles and calculates the average value $\mathbf{\Psi}_{\mathrm{t}}$ at the 2 vertices and face point for each triangle, area $|\mathbf{S}_{\mathrm{t}}|$. Point interpolation calculates the face value as the area-weighted average of triangle values, *i.e.* $\mathbf{\Psi}_{\mathrm{f}} = (\sum |\mathbf{S}_{\mathrm{t}}| \mathbf{\Psi}_{\mathrm{t}}) / |\mathbf{S}_{\mathrm{f}}|$.

## Least squares gradient

A gradient calculation using a **least squares** finite difference method is sometimes used within the finite volume framework. The method calculates the gradient in a cell which, when used to extrapolate the cell value to centres of all neighbouring cells, minimises the error between extrapolated values and cell values.

For a given cell, a tensor $\mathbf{G}$ is calculated by summing over faces using the inverse distance weighting $w_{\mathrm{id}} = 1/|\mathbf{\Delta d}|$, where $\mathbf{\Delta d}$ is the cell centre-centre vector:

$$\mathbf{G} = \sum_{\mathrm{f}} w_{\mathrm{id}}^2 \mathbf{\Delta d} \mathbf{\Delta d}. \tag{3.19}$$

The gradient is then evaluated using the inverse of the tensor $\mathbf{G}^{-1}$ and values in the neighbour (N) and current (P) cells:

$$\nabla \mathbf{\Psi} \rightarrow \frac{1}{V} \sum_{\mathrm{f}} w_{\mathrm{id}}^2 \mathbf{G}^{-1} \bullet \mathbf{\Delta d} \left( \mathbf{\Psi}_{\mathrm{N}} - \mathbf{\Psi}_{\mathrm{P}} \right). \tag{3.20}$$

## 3.16 Gradient limiting

The standard gradient calculation can generate values with an unphysically large magnitude when cell quality is poor. A large gradient value in one cell can contribute sufficiently to the source term of an equation to cause unboundedness in the solution.

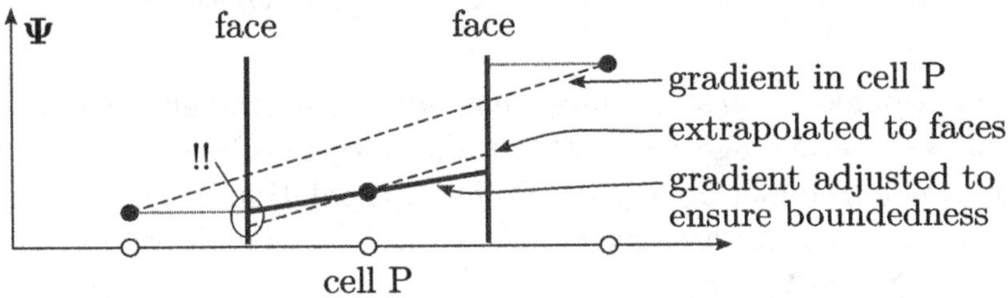


The calculated gradient can be *limited* to avoid unphysical values. The limiting extrapolates the cell values $\boldsymbol{\Psi}$ using the gradient to *all* neighbouring face centres. If the extrapolated $\boldsymbol{\Psi}_{\mathrm{f}}$ at a face falls outside of the bounds of $\boldsymbol{\Psi}$ at neighbouring cell centres, the gradient is adjusted to match the bounding value.

This numerical scheme has an advantage that is takes into account all surrounding cells so is fully "multi-dimensional". It ensures boundedness locally between cells so can help eliminate instabilities in *advection* discretisation.

### Linear upwind with gradient limiting

Limiting can be applied to the cell gradient in the explicit contribution of the linear upwind advection scheme, described in Sec. 3.14. The scheme with gradient limiting is analysed below for its TVD behaviour on a regular mesh in 1D.

In the following figure, the cell gradient is denoted by "$\nabla$" and the face gradient is denoted by "$\nabla_{\mathrm{f}}$". The ratio of gradients is $r = \max\left[2(\nabla/\nabla_{\mathrm{f}}) - 1, 0\right]$ from Eq. (3.12), which is adjusted by modifying $\Psi_P$. No limiting is applied for $1/3 \leq r \leq 3$, when $\Psi_f$ can be expressed by $\Psi_f = \Psi_P + \beta(\Psi_L - \Psi_P)$, with $\beta = \nabla/\nabla_f$. In that case, $r = 2\beta - 1$, corresponding to a function $\beta = (r+1)/2$.

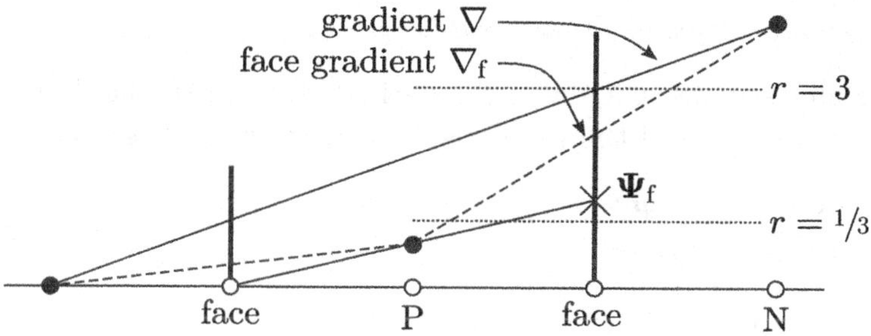


Limiting is applied for $r < 1/3$, reducing the gradient linearly to zero as $r \rightarrow 0$. This corresponds to a function $\beta = 2r$ for $0 < r < 1/3$. Similarly, limiting is applied when $r > 3$. Within that range, $\Psi_f = \Psi_N$ so $\beta = 2$.

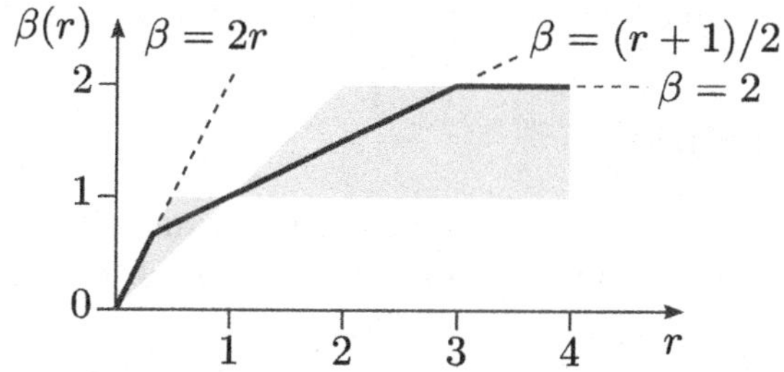


On a Sweby diagram linear upwind with gradient limiting is TVD, with the line $\beta = (r+1)/2$ being the regime with no gradient limiting.

This function uses too much downwind interpolation to be a reliable TVD scheme for advection. However, as a gradient limiter for the linear upwind scheme, it is strongly bounded, due to the multi-dimensional nature of gradient limiting — with good accuracy due to the skewness correction in linear upwind.

## 3.17 Time discretisation

A local time derivative $\partial\boldsymbol{\Psi}/\partial t$ in an equation can be discretised as a finite difference in time. Time $t$ is expressed in discrete intervals, or *steps*, of duration $\Delta t$.

The **Euler** scheme calculates the derivative from the field at the current time $\boldsymbol{\Psi}$ and the previous, or *old*, time level $\boldsymbol{\Psi}^{\mathrm{o}}$ by:

$$\frac{\partial\boldsymbol{\Psi}}{\partial t} \rightarrow \frac{\boldsymbol{\Psi} - \boldsymbol{\Psi}^{\mathrm{o}}}{\Delta t} \tag{3.21}$$

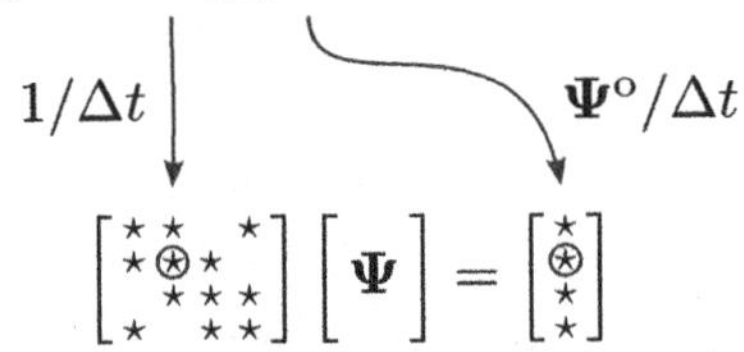


Time discretisation thereby contributes to the *diagonal* (*only*!) coefficients of the matrix $[\mathbf{A}]$ and to the source vector $[\mathbf{b}]$.

### Courant number

For a 1D domain in the $x$-direction, the Courant number is the following dimensionless parameter for each cell of length $\Delta x$:

$$\mathrm{Co} = \frac{u_x \Delta t}{\Delta x}. \tag{3.22}$$

The Courant number originates from the solution of 1D advection, *e.g.* Eq. (2.32), with the Euler time scheme of Eq. (3.21) and an *explicit* upwind advection scheme from Sec. 3.10. For that case, the Courant-Friedrichs-Lewy (CFL) condition† for convergence is $\mathrm{Co} \leq 1$ across *all cells*. $\mathrm{Co} = 1$ corresponds

†after Richard Courant, Kurt Friedrichs and Hans Lewy, *Uber die partiellen Differenzengleichungen der mathematischen Physik*, 1928.

to a fluid particle moving across one cell in one time step, so its relevance to solution convergence is perhaps unsurprising.

In an *explicit* solution, the convergence limit can reduce further to $\mathrm{Co} \leq 0.5$ or $\mathrm{Co} \leq 0.25$, with more accurate schemes for advection. But the finite volume method is generally *implicit* so stability can then be maintained with a higher maximum value of Co. Temporal accuracy is then the important consideration when choosing the Co of a simulation, both in terms of its mean and maximum value across all cells in the domain.

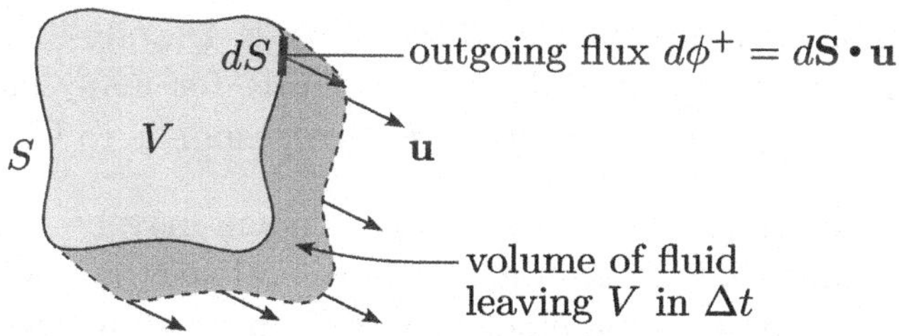


It is therefore important to monitor Courant number which needs to be calculated for 3D problems. Explicit, Euler, upwind discretisation of advection can be presented in 3D as follows:

$$\mathbf{\Psi} = \mathbf{\Psi}^{\mathrm{o}} - \frac{\Delta t}{V}\sum_{\mathrm{f}} \phi^{+}\mathbf{\Psi}^{\mathrm{o}} - \frac{\Delta t}{V}\sum_{\mathrm{f}} \phi^{-}\mathbf{\Psi}^{\mathrm{o}}_{\mathrm{N}} = 0. \tag{3.23}$$

Here $\phi^+$ are positive fluxes which transports $\mathbf{\Psi}^{\mathrm{o}}$ out of the cell of interest and $\phi^-$ are negative fluxes which transports $\mathbf{\Psi}^{\mathrm{o}}_{\mathrm{N}}$ from neighbouring cells. Since the CFL condition, $\mathrm{Co} \leq 1$, requires that the coefficient of $\mathbf{\Psi}^{\mathrm{o}}$ cannot be negative, it follows that

$$\mathrm{Co} = \frac{\Delta t}{V}\sum_{\mathrm{f}} \phi^{+}. \tag{3.24}$$

This 3D Co represents the volume of fluid leaving the cell in one time step, as a fraction of the cell volume, as shown above.

## 3.18 Second order time schemes

A Taylor's series expansion between the current time $t$ and 'old' time at $t - \Delta t$ relates to the Euler *implicit* scheme by

$$\frac{\boldsymbol{\Psi} - \boldsymbol{\Psi}^{\mathrm{o}}}{\Delta t} = \frac{\partial \boldsymbol{\Psi}}{\partial t} - \frac{\partial^2 \boldsymbol{\Psi}}{\partial t^2}\frac{\Delta t}{2} + \mathcal{O}(\Delta t^2)\ldots \tag{3.25}$$

The truncation error $\propto \Delta t^1$, *i.e.* it is *first order accurate* in time when the time derivative relates to $\boldsymbol{\Psi}$ *at the current time*, as shown in the figure on the following page. Despite its low order, the Euler scheme is sufficiently accurate for many simulations since $\Delta t$ is generally small when it corresponds to $\mathrm{Co} \approx 1$.

Nevertheless, a second order time scheme may be required for simulations that demand higher temporal accuracy or to enable greater computational efficiency by running with larger $\Delta t$.

### Backward scheme

In Eq. (3.25) we can replace $\boldsymbol{\Psi}^{\mathrm{o}}$ by the values $\boldsymbol{\Psi}^{\mathrm{oo}}$ at 'old-old' time $t - 2\Delta t$. Subtracting the expression from Eq. (3.25) and rearranging terms gives the following relation for the *second* derivative

$$\frac{\partial^2 \boldsymbol{\Psi}}{\partial t^2} = \frac{\boldsymbol{\Psi} - 2\boldsymbol{\Psi}^{\mathrm{o}} + \boldsymbol{\Psi}^{\mathrm{oo}}}{\Delta t^2} + \mathcal{O}(\Delta t^2)\ldots \tag{3.26}$$

Substituting Eq. (3.26) into Eq. (3.25) gives the **backward** scheme which is second order accurate, using values at three time levels $\boldsymbol{\Psi}$, $\boldsymbol{\Psi}^{\mathrm{o}}$ and $\boldsymbol{\Psi}^{\mathrm{oo}}$:

$$\frac{\partial \boldsymbol{\Psi}}{\partial t} \rightarrow \frac{3\boldsymbol{\Psi} - 4\boldsymbol{\Psi}^{\mathrm{o}} + \boldsymbol{\Psi}^{\mathrm{oo}}}{2\Delta t} \tag{3.27}$$

## Crank-Nicolson scheme

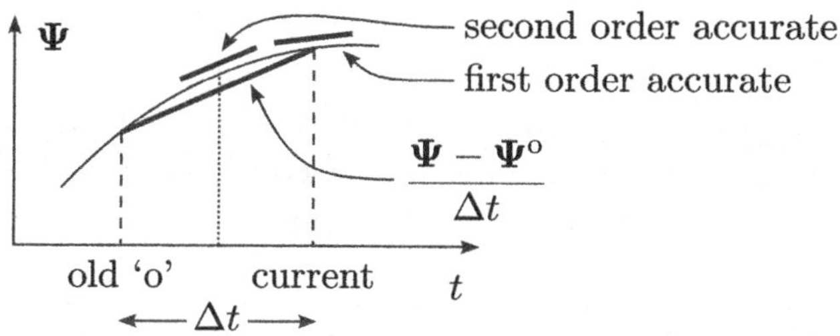


An implicit solution expresses the terms in an equation, *e.g.* advection, Laplacian, at the *current* time. The **Crank-Nicolson** method,† expresses the terms at the *midpoint* between the current and old times, to make the Euler time scheme second order accurate. Denoting discretised terms except the time derivative by $[\mathbf{A}\,|\,\mathbf{b}]\,\boldsymbol{\Psi}$, the Crank-Nicolson method solves

$$\left.\frac{\partial \boldsymbol{\Psi}}{\partial t}\right|_{\text{Euler}} + \frac{1}{2}[\mathbf{A}\,|\,\mathbf{b}]\,\boldsymbol{\Psi} + \frac{1}{2}[\mathbf{A}\,|\,\mathbf{b}]^{\text{o}}\,\boldsymbol{\Psi}^{\text{o}} = 0, \tag{3.28}$$

where $[\mathbf{A}\,|\,\mathbf{b}]^{\text{o}}\,\boldsymbol{\Psi}^{\text{o}}$ is *calculated* using old time values $\boldsymbol{\Psi}^{\text{o}}$.

A modern version of the scheme replaces the two $1/2$ factors by $1/(1+\lambda)$ and $\lambda/(1+\lambda)$, introducing the 'offset coefficient' $\lambda$, where $\lambda = 0$ corresponds to Euler implicit and $\lambda = 1$ is Crank-Nicolson Eq. (3.28). If $[\mathbf{A}\,|\,\mathbf{b}]\,\boldsymbol{\Psi}$ is discretised *implicitly* (as normal), the Crank-Nicolson scheme can be represented as a time derivative discretised by

$$\frac{\partial \boldsymbol{\Psi}}{\partial t} \rightarrow (1+\lambda)\frac{\boldsymbol{\Psi} - \boldsymbol{\Psi}^{\text{o}}}{\Delta t} + \lambda\,[\mathbf{A}\,|\,\mathbf{b}]^{\text{o}}\,\boldsymbol{\Psi}^{\text{o}}. \tag{3.29}$$

In practice, $\lambda = 0.9$ is generally used to ensure solution stability.

† John Crank and Phyllis Nicolson, *A practical method for numerical evaluation of solutions of partial differential equations of the heat conduction type*, 1947.

## 3.19 Calculated derivatives

The introduction to matrix construction, Sec. 3.5, described implicit and explicit discretisation of terms in an equation. It concluded that the principal derivatives of $\mathbf{\Psi}$ that can be treated implicitly — forming matrix coefficients in $[\mathbf{A}]$ — are a time derivative, advection and Laplacian.

Terms with other derivatives must be calculated from respective fields, *e.g.* $\nabla p$ from current values of $p$. In Sec. 3.15, we have described discretisation of a gradient, which is always explicit. This section gathers together the other derivatives found in equations for fluid dynamics and associated models.

### General divergence term

A general divergence term is any term that can be represented by $\nabla \bullet \mathbf{\Psi}$. It excludes the Laplacian term which includes a gradient $\nabla\mathbf{\Psi}$, and advection which includes $\mathbf{u}$.

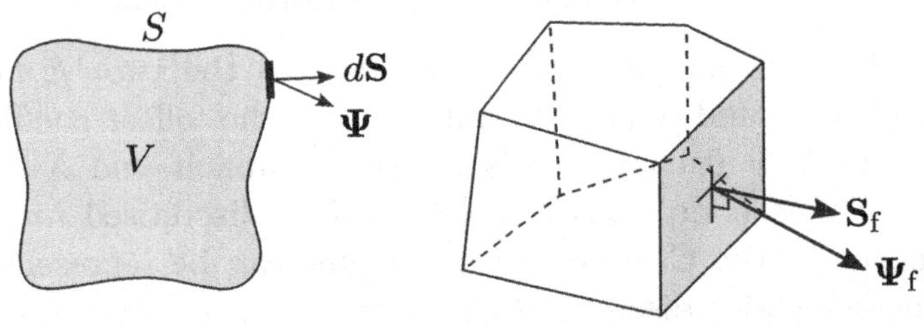


The discretisation of a general divergence term is an explicit calculation using current values of $\mathbf{\Psi}$. It is based on a surface integral using the divergence definition in Sec. 2.23 as shown below:

$$\nabla \bullet \mathbf{\Psi} = \lim_{\Delta V \to 0} \frac{1}{\Delta V} \int_S (d\mathbf{S} \bullet \mathbf{\Psi}) \to \frac{1}{V} \sum_{\mathrm{f}} \mathbf{S}_{\mathrm{f}} \bullet \mathbf{\Psi}_{\mathrm{f}}. \qquad (3.30)$$

The face value $\boldsymbol{\Psi}_{\mathrm{f}}$ is generally interpolated from cell values using the linear scheme. Terms discretised using this scheme include $\nabla \bullet \left[\mu(\nabla\mathbf{u})^{\mathrm{T}}\right]$ in Eq. (2.45), a divergence of stress $\nabla \bullet \boldsymbol{\sigma}$, *etc.*

## Curl of a vector

The curl derivative $\nabla \times \mathbf{u}$ is calculated from the gradient $\nabla\mathbf{u}$ and applying the Hodge dual operator given by Eq. (2.40) using the following relation:

$$\nabla \times \mathbf{u} = 2 *(\operatorname{skw} \nabla\mathbf{u}) \,. \tag{3.31}$$

In other words, $\nabla\mathbf{u}$ is discretised according to a scheme from Sec. 3.15, from which $\nabla \times \mathbf{u}$ is calculated by Eq. (3.31).

## Mag-square grad-grad

A derivative which appears in some model equations is $|\nabla\nabla\boldsymbol{\Psi}|^2$, described as “mag-square grad-grad”. This derivative returns a scalar since the mag-square, *e.g.* $|\mathbf{a}|^2$, represents the inner product of $\mathbf{a}$ with itself, as shown in Eq. (2.7).

The mag-square calculation always uses the appropriate inner product to reduce the result to a scalar. For a tensor $\mathbf{T}$, it is the double inner product, *i.e.* $|\mathbf{T}|^2 = \mathbf{T} \mathbin{:} \mathbf{T}$.

The grad-grad operator $\nabla\nabla$ yields a third-rank tensor in the case that $\boldsymbol{\Psi}$ is a vector field. To avoid storing third-rank tensors, the mag-square grad-grad operator is evaluated by summing the result from the operator on each *component* $i$ of $\boldsymbol{\Psi}$ by

$$\sum_{i=0}^{N-1} |\nabla\nabla\boldsymbol{\Psi}_i|^2 \,, \tag{3.32}$$

where $N$ is the number of components in $\boldsymbol{\Psi}$.

## 3.20 Other terms

In an equation for $\boldsymbol{\Psi}$, there can be terms other than the time derivative, advection and Laplacian of $\boldsymbol{\Psi}$, including:

- a linear function $c\boldsymbol{\Psi}$ where $c$ is a scalar coefficient or field;
- a term without $\boldsymbol{\Psi}$, sometimes including the derivative of another variable, *e.g.* $\nabla p$.

The second example is simply discretised as an *explicit* gradient term as described in Sec. 3.15. Derivative terms like this one are described in earlier sections and require no further discussion.

The first example, the $c\boldsymbol{\Psi}$ term, requires much more discussion, particularly relating to the possibility of *implicit* discretisation. Let us consider the following equation:

$$\frac{\partial \boldsymbol{\Psi}}{\partial t} - c\boldsymbol{\Psi} = 0\,. \tag{3.33}$$

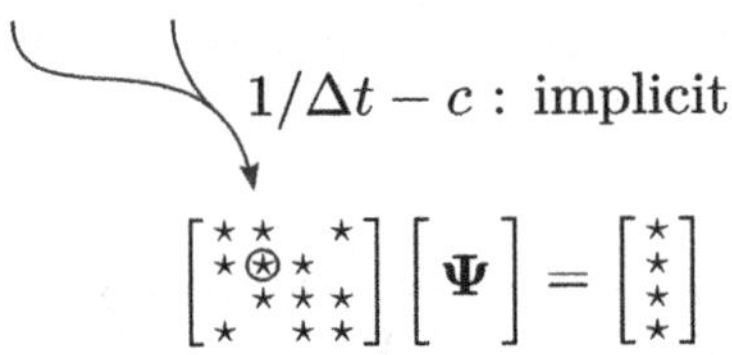


If $c\boldsymbol{\Psi}$ is discretised implicitly, then the matrix would contain a zero diagonal coefficient when $c = 1/\Delta t$, making it *singular* or *non-invertible.* If this occurs, $\boldsymbol{\Psi}$ is not present in the linear equation for the relevant cell, so it cannot be solved.

Therefore, such a term *must* be discretised *explicitly* when it has a *negative sign* (or positive on the right side of "="), to ensure the matrix equation is solvable. The nature of Eq. (3.33) is that $\boldsymbol{\Psi}$ can only increase from an initial positive value.

## Implicit discretisation of linear terms

Let us now consider the equivalent equation with a linear term that has a positive sign, *i.e.*

$$\frac{\partial \boldsymbol{\Psi}}{\partial t} + c\boldsymbol{\Psi} = 0\,. \tag{3.34}$$

With Eq. (3.34), $\boldsymbol{\Psi}$ can only decrease from an initial positive value but reaches a lower limit at $\boldsymbol{\Psi} = 0$. Like many scalar properties, *e.g.* $\rho$, $T$, $e$, *etc.*, $\boldsymbol{\Psi}$ may have a lower physical bound of 0, which the equation intentionally reflects.

It is important to maintain a physical bound of $\boldsymbol{\Psi}$. Discretisation of Eq. (3.34) using the Euler time scheme Eq. (3.21) gives

$$\boldsymbol{\Psi} = \frac{1}{1 + c\Delta t}\boldsymbol{\Psi}^{\mathrm{o}}, \tag{3.35}$$

which ensures boundedness since $\boldsymbol{\Psi} > 0$. An equivalent explicit discretisation gives $\boldsymbol{\Psi} = (1 - c\Delta t)\boldsymbol{\Psi}^{\mathrm{o}}$ which is only bounded when $c \leq 1/\Delta t$, similar to the Co limit imposed by explicit discretisation of advection, described in Sec. 3.17. To avoid this $\Delta t$ limit, we apply *implicit* discretisation to terms with a *positive sign* on the left side of "=", where possible.

Even when the term is not linear in $\boldsymbol{\Psi}$, it can be implemented as such by "dividing and multiplying by $\boldsymbol{\Psi}$". For example, the $k$–$\varepsilon$ turbulence model includes an $\varepsilon$ term in the $k$ equation, *i.e.* $\partial k/\partial t + \varepsilon \ldots$, ignoring other terms. Dividing and multiplying the $\varepsilon$ term by $k$ gives

$$\frac{\partial k}{\partial t} + \left(\frac{\varepsilon}{k}\right) k \ldots \tag{3.36}$$

producing a linear term in $k$ which can be discretised implicitly using a coefficient $c = \varepsilon/k$.

## 3.21 Terms which change sign

An equation may include a function $f$ which can return a value which is positive or negative, *e.g.*

$$\frac{\partial \boldsymbol{\Psi}}{\partial t} + f = 0. \tag{3.37}$$

Following Sec. 3.20, we aim to avoid a singular matrix in the case $f < 0$ and ensure the solution exceeds the lower bound of 0 when $f \geq 0$. The discretisation of $f$ is therefore treated implicitly or explicitly within each cell based on

$$f = \begin{cases} \left(\dfrac{f}{\boldsymbol{\Psi}}\right) \boldsymbol{\Psi}, & \text{if } f \geq 0; \\ f(\text{explicit}), & \text{otherwise.} \end{cases} \tag{3.38}$$

### Maintaining lower and upper bounds

The solution variable $\boldsymbol{\Psi}$ could have a physical bound at any low value $\boldsymbol{\Psi}_{\mathrm{L}}$ and/or high value $\boldsymbol{\Psi}_{\mathrm{H}}$. We can treat the function $f$ in Eq. (3.37) as follows to maintain boundedness:

$$f = \begin{cases} f_{\mathrm{L}}(\boldsymbol{\Psi} - \boldsymbol{\Psi}_{\mathrm{L}}), & \text{if } f \geq 0; \\ f_{\mathrm{H}}(\boldsymbol{\Psi}_{\mathrm{H}} - \boldsymbol{\Psi}), & \text{otherwise.} \end{cases} \tag{3.39}$$

Here, $f_{\mathrm{L}} = f/(\boldsymbol{\Psi} - \boldsymbol{\Psi}_{\mathrm{L}})$ and $f_{\mathrm{H}} = f/(\boldsymbol{\Psi}_{\mathrm{H}} - \boldsymbol{\Psi})$ are calculated using the current values of $\boldsymbol{\Psi}$.

If we examine the situation with $f \geq 0$, Eq. (3.37) would be discretised as

$$\frac{\partial \boldsymbol{\Psi}}{\partial t} + f_{\mathrm{L}} \boldsymbol{\Psi} = f_{\mathrm{L}} \boldsymbol{\Psi}_{\mathrm{L}}. \tag{3.40}$$

When we discretise $f_{\mathrm{L}}\boldsymbol{\Psi}$ implicitly alongside the Euler time scheme, Eq. (3.40) becomes:

$$\left(\frac{1}{\Delta t} + f_{\mathrm{L}}\right)\boldsymbol{\Psi} = \left(\frac{1}{\Delta t} + \frac{\boldsymbol{\Psi}_{\mathrm{L}}}{\boldsymbol{\Psi}^{\mathrm{o}}} f_{\mathrm{L}}\right)\boldsymbol{\Psi}^{\mathrm{o}}. \tag{3.41}$$

The solution for $\boldsymbol{\Psi}$ decreases when $\boldsymbol{\Psi}^{\mathrm{o}} > \boldsymbol{\Psi}_{\mathrm{L}}$, but cannot fall below the $\boldsymbol{\Psi}_{\mathrm{L}}$ bound within a solution step. In the limit that $\boldsymbol{\Psi}^{\mathrm{o}} \rightarrow \boldsymbol{\Psi}_{\mathrm{L}}$, the solution for $\boldsymbol{\Psi}$ no longer decreases.

An equivalent analysis shows Eq. (3.40) maintains boundedness for the upper bound $\boldsymbol{\Psi}_{\mathrm{H}}$. In that case, note that $f$ is negative so the term in $\boldsymbol{\Psi}$ can still be treated implicitly.

## Fields bounded by 0 and 1

Many properties are expressed as a fraction, *e.g.* concentration of a chemical species, so are bounded by 0 and 1. When that happens, Eq. (3.39) simplifies to

$$f = \begin{cases} f_{\mathrm{L}}\boldsymbol{\Psi}, & \text{if } f \geq 0; \\ f_{\mathrm{H}}(1 - \boldsymbol{\Psi}), & \text{otherwise.} \end{cases} \tag{3.42}$$

Here, $f_{\mathrm{L}} = f/\boldsymbol{\Psi}$ and $f_{\mathrm{H}} = f/(1 - \boldsymbol{\Psi})$ are calculated using the current values of $\boldsymbol{\Psi}$.

It is particularly important to maintain boundedness of a field with 0-1 bounds when it is used to calculate another property which has a physical bound. For example, in a system of $i$ fluid phases, fluid $\rho$ can be calculated by $\rho = \sum_i \alpha_i \rho_i$ from phase fractions $\alpha_i$ and densities $\rho_i$.

A small amount of unboundedness in $\alpha_i$, in the phase with highest $\rho_i$, causes large unboundedness in $\rho$, *e.g.* in 2 phases with $\rho_1 = 1000$ and $\rho_2 = 1$, the calculated $\rho \approx -9$ with $\alpha_1 = -0.01$.

## 3.22 Bounded advection discretisation

Sec. 2.9 described the conservation and boundedness qualities of the advective derivatives $\nabla \bullet (\mathbf{u}\boldsymbol{\Psi})$ and $\mathbf{u} \bullet \nabla\boldsymbol{\Psi}$, respectively. According to Eq. (2.31), the terms are equivalent when $\nabla \bullet \mathbf{u} = 0$, *i.e.* when mass conservation Eq. (2.8) is satisfied in the case when $\rho =$ constant.

During computation of an incompressible flow, numerical error can be significant, in particularly at the beginning of a steady-state calculation. The governing equations are not satisfied, including mass conservation, so $\nabla \bullet \mathbf{u} \neq 0$.

Discretisation of an advection term of the form $\nabla \bullet (\mathbf{u}\boldsymbol{\Psi})$ as described in Sec. 3.9 can cause unboundedness in $\boldsymbol{\Psi}$. If $\boldsymbol{\Psi}$ has a physical bound that is violated, the solution rapidly becomes unstable.

The addition of the $-(\nabla \bullet \mathbf{u})\boldsymbol{\Psi}$ term ensures boundedness while $\nabla \bullet \mathbf{u} \neq 0$, at the expense of conservation. Once $\nabla \bullet \mathbf{u} = 0$, the solution is both bounded and conservative.

$$\mathbf{u} \bullet \nabla\boldsymbol{\Psi} \rightarrow \nabla \bullet (\mathbf{u}\boldsymbol{\Psi}) - (\nabla \bullet \mathbf{u})\boldsymbol{\Psi}. \tag{3.43}$$

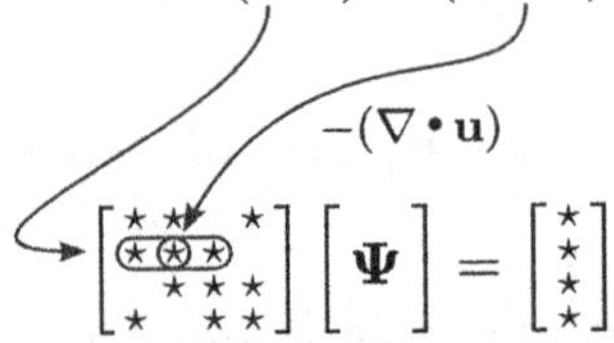


This additional term is implicit in $\boldsymbol{\Psi}$, so contributes $-\nabla \bullet \mathbf{u}$ to the diagonal coefficients of the matrix. If $\nabla \bullet \mathbf{u}$ is positive, we might expect a singular matrix, as discussed in Sec. 3.20. However, the decrease to the diagonal coefficient is matched by an increase from the discretisation of $\nabla \bullet (\mathbf{u}\boldsymbol{\Psi})$. If that term uses the upwind scheme, the discretisation can be split into a contribution from positive outgoing fluxes $\phi^{+}$ and negative incoming fluxes $\phi^-$. In extensive form (*i.e.* scaled by $V$, see Sec. 3.24), using values in the cell of interest $\boldsymbol{\Psi}$ and neighbouring cells $\boldsymbol{\Psi}_\mathrm{N}$, the terms are

$$\underbrace{\sum_\mathrm{f} \phi^+ \boldsymbol{\Psi} + \sum_\mathrm{f} \phi^- \boldsymbol{\Psi}_\mathrm{N}}_{\nabla \bullet (\mathbf{u}\boldsymbol{\Psi})} - \underbrace{\sum_\mathrm{f} \phi^+ \boldsymbol{\Psi} + \sum_\mathrm{f} \phi^- \boldsymbol{\Psi}}_{(\nabla \bullet \mathbf{u})\boldsymbol{\Psi}}.$$

The $\sum_\mathrm{f} \phi^+ \boldsymbol{\Psi}$ terms cancel, leaving negative coefficients for neighbouring cells (since $\phi^-$ is negative). The diagonal coefficient is then equal to the sum of magnitude of those neighbour cell coefficients, resulting in an invertible matrix.

## Capturing physics

Boundedness and conservation can be compromised when a term in an equation does not capture the physics of the problem. An example from flame speed combustion modelling uses a parameter $b$ which represents the fraction of the unburnt fuel mixture.

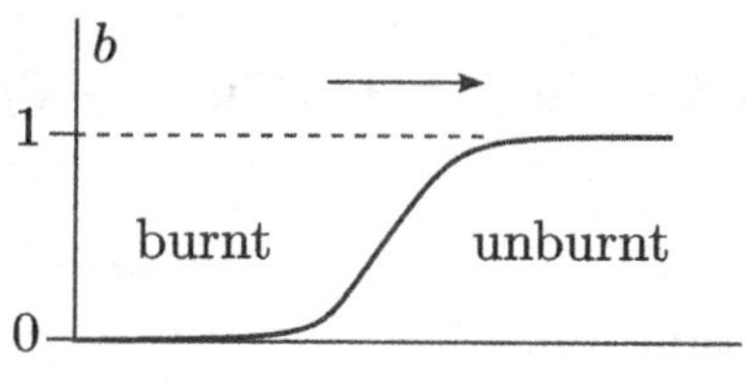


The equation for $b$ includes the source term $u_b|\nabla b|$, where $u_b$ is a calculated flame speed. Its inclusion could cause the solution of $b$ to fall below its lower bound of 0.

Multiplying and dividing by $|\nabla b| = \nabla b \bullet \nabla b$ changes the term to $\mathbf{u}_b \bullet \nabla b$, where $\mathbf{u}_b = u_b \mathbf{n}_b$ and unit vector $\mathbf{n}_b = \nabla b / |\nabla b|$. In this form, the term represents the non-conservative advection of $b$ by a flame velocity $\mathbf{u}_b$ moving in the direction $\mathbf{n}_b$, which captures the physical nature of combustion. Boundedness can then be maintained by a suitable choice of advection scheme.

## 3.23 Recommended discretisation schemes

The following figures provide a summary and recommendations of discretisation schemes for different equations. We start with momentum conservation Eq. (2.49), described in Sec. 2.13 and Sec. 2.14, for which *accuracy* is the main consideration in scheme selection.

**Time**, Sec. 3.17
Usually *1st-order*
Euler scheme, Eq. (3.21)

*2nd-order* to preserve
transient structures,
*e.g.* large eddies, waves
Crank-Nicolson, Eq. (3.29)
backward implicit, Eq. (3.27)

**Laplacian**, Sec. 3.7
*surface normal*
*gradient*, Sec. 3.8

non-orthogonal correction
Eq. (3.7) for $\theta_{\text{no}} \lesssim 75^\circ$

limited correction
for $\theta_{\text{no}} > 75^\circ$

$$\frac{\partial \mathbf{u}}{\partial t} + \nabla \bullet (\mathbf{u}\mathbf{u}) - \nabla \bullet (\nu \nabla \mathbf{u}) = -\nabla p$$

**Advection**, Sec. 3.9
linear interpolation, Eq. (3.4)
creeping flows, large eddy simulation

linear upwind, Sec. 3.14
with gradient limiting, Sec. 3.16

limited linear, Eq. (3.13)
minmod, Eq. (3.14)

upwind interpolation, Sec. 3.10
fast-converging, approximate solution

*decreasing accuracy*
*increasing stability*

**Gradient**, Eq. (3.18)

## Equation for a bounded property

The discretisation of an equation for a bounded scalar field is illustrated using the $k$ equation from the $k$–$\varepsilon$ turbulence model (assuming constant $\rho$). Schemes are selected with an emphasis on *boundedness*.

**Time**, Sec. 3.17
see momentum
equation example

**Turbulence generation**
calculated gradient, Sec. 3.15

**Laplacian**, Sec. 3.7
see momentum
equation example

**Linear in $k$**
switches between implicit
and explicit, Eq. (3.38)

$$\frac{\partial k}{\partial t} + \nabla \bullet (\mathbf{u}k) - \nabla \bullet (D_{\text{eff}} \nabla k) = G(\nabla \mathbf{u}) - \frac{2}{3}(\nabla \bullet \mathbf{u})k - \varepsilon$$

**Advection**, Sec. 3.9
linear upwind, Sec. 3.14
with gradient limiting, Sec. 3.16

**Other term**, Sec. 3.20
made implicit in $k$ by
"divide and multiply"
$(\varepsilon/k)\,k$, Eq. (3.36)

*limited schemes*, Sec. 3.11
van Leer, Eq. (3.15)
van Albada, Eq. (3.16)
limited linear, Eq. (3.13)
minmod, Eq. (3.14)

*decreasing accuracy*

*increasing stability*

upwind interpolation, Sec. 3.10

## 3.24 Example of building a matrix equation

The previous sections describe methods to discretise derivative and other terms in order to build a matrix equation for a given physical equation. Let us demonstrate the construction of a matrix equation, using the momentum conservation equation from Sec. 3.23 as an example. It is a vector equation, so produces 3 matrix equations for $[\,u_x\,]$, $[\,u_y\,]$ and $[\,u_z\,]$.

The first term, the time derivative $\partial \mathbf{u}/\partial t$, might be discretised with the Euler scheme Eq. (3.21). Matrix equations are constructed in *extensive form* as discussed in Sec. 3.6. Hence, the contributions from Eq. (3.21) to matrix coefficients $[\,\mathbf{A}\,]$ and source vector $[\mathbf{b}]$ are scaled by cell volume $V$, *i.e.* $V/\Delta t$ and $(V/\Delta t)\mathbf{\Psi}^{\text{o}}$, respectively, as illustrated below.

$$\frac{\partial \mathbf{u}}{\partial t} + \nabla \bullet (\mathbf{u}\mathbf{u}) - \nabla \bullet (\nu \nabla \mathbf{u}) = -\nabla p$$

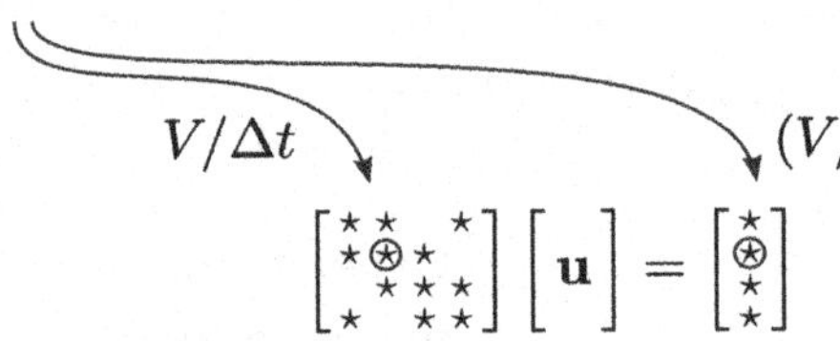


The second term, the advective derivative $\nabla \bullet (\mathbf{u}\mathbf{u})$, is discretised by Eq. (3.8). It pre-calculates the volumetric flux $\phi_{\text{f}} = \mathbf{S}_{\text{f}} \bullet \mathbf{u}_{\text{f}}$, using $\mathbf{u}_{\text{f}}$ interpolated by Eq. (3.3) with linear weights Eq. (3.4).

The transported $\mathbf{u}_{\text{f}}$ might be discretised using the linear upwind scheme described in Sec. 3.14. The scheme first applies upwind discretisation, which contributes outgoing positive fluxes $\phi^{+}$ to diagonal coefficients and negative fluxes $\phi^{+}$ to off-diagonals. It then adds an explicit contribution based on an extrapolated gradient $\mathbf{d}_{\text{P}} \bullet \nabla \mathbf{u}$ (see Sec. 3.14). The gradient $\nabla \mathbf{u}$ is usually calculated by Eq. (3.18) with gradient limiting from Sec. 3.16.

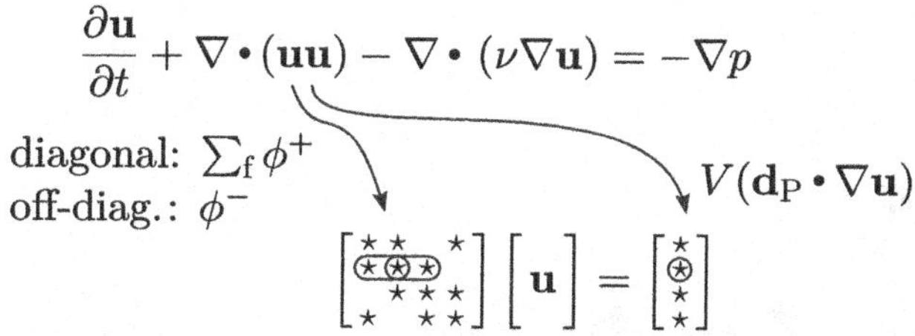


The third term, the Laplacian derivative $-\nabla \cdot (\nu \nabla \mathbf{u})$, is discretised by Eq. (3.2). It requires $\nu_{\mathrm{f}}$, which is linearly interpolated from the cell centres. If the surface normal gradient $\nabla_{\mathbf{n}} \mathbf{u}_{\mathrm{f}}$ includes a non-orthogonal correction $\nabla_{\mathbf{n}} \boldsymbol{\Psi}_{\mathrm{f}}^{\mathrm{corr}}$, see Sec. 3.8, then the term contributes to $[\,\mathbf{A}\,]$ and $[\mathbf{b}]$, as shown below.

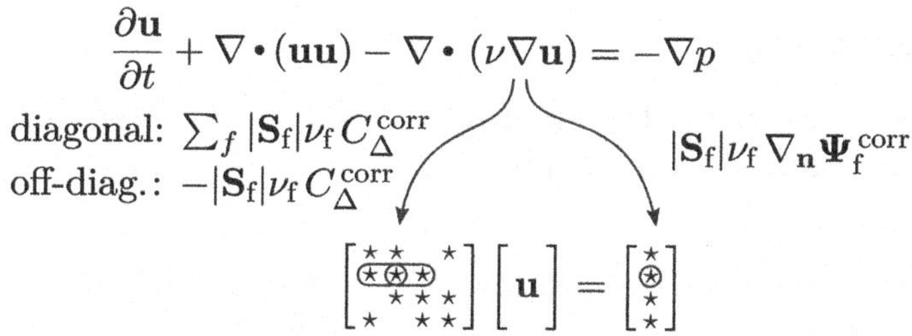


The final term, $-\nabla p$ is calculated using Eq. (3.18). Like all the other terms described here, it is implemented in extensive form, scaling by $V$, so is calculated for each cell by the vector $\sum_{\mathrm{f}} \mathbf{S}_{\mathrm{f}} p$. The relevant component ($x$, $y$, $z$) of this vector is then applied to the respective equation for $[\,u_x\,]$, $[\,u_y\,]$ and $[\,u_z\,]$.

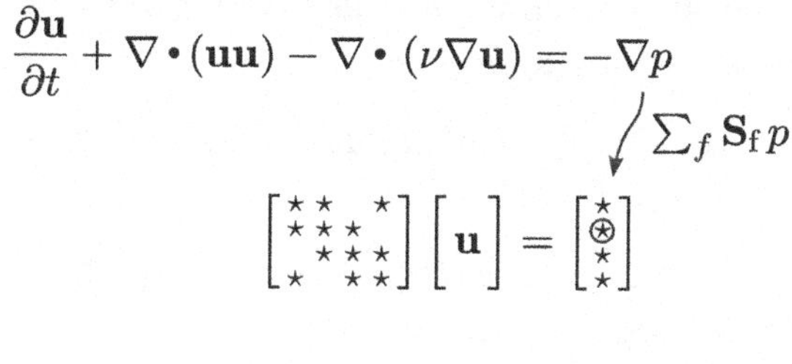

# Chapter 4

# Boundary Conditions

*"Sprawling on the fringes of the city, in geometric order, an insulated border."*

Rush, *Subdivisions* (1982).

Chapter 3 described:

- the computational mesh consisting of cells of any irregular polyhedral shape with mesh properties for the finite volume method;
- discretisation of terms in equations to construct a matrix equation $[\mathbf{A}][\boldsymbol{\Psi}] = [\mathbf{b}]$ for solution variable $\boldsymbol{\Psi}$;
- discretisation schemes for optimal accuracy, boundedness and convergence, using limiting and correction;
- the time derivative, Courant number and time step size.

At the boundary of the mesh, constraints must be applied to the solution variable $\boldsymbol{\Psi}$, known as *boundary conditions*. Setting boundary conditions is challenging because:

- they need to *reflect the physical conditions* of the case being simulated;
- they need to make the case *well posed*, *i.e.* provide unique, stable, physical solutions to the equations;

- they must be compatible across sets of multiple equations, in particular $p$ and $\mathbf{u}$.

The mesh boundary is split into regions known as *patches*, on which different boundary conditions are applied. The choice of boundary condition generally depends on flow direction at a patch, whether the patch corresponds to a solid wall, *etc.*

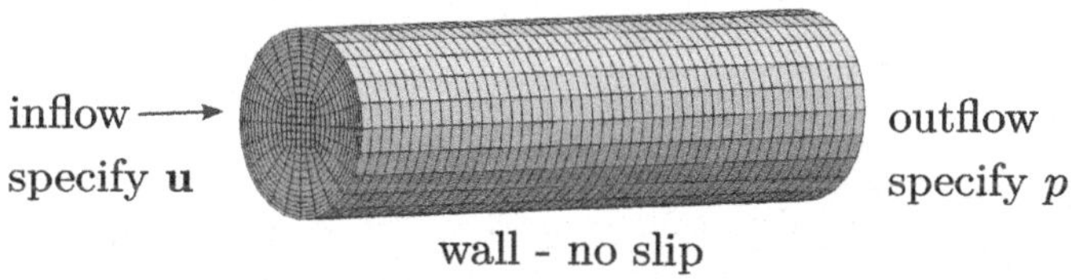


There are basic forms of boundary condition which specify the value, normal gradient, *etc.* of $\boldsymbol{\Psi}$ at the boundary. They are applied through modifications to the matrix coefficients $[\mathbf{A}]$ and source vector $[\mathbf{b}]$ using mesh data of the faces and cells adjacent to the boundary.

The opening topics of this chapter describe the mesh data and the numerical methods required by the basic forms of boundary condition.

More specialised boundary conditions can be *derived* from the underlying forms which correspond to different physical conditions. Some derived conditions, that are often used for particular boundary configurations, are introduced in this chapter.

Otherwise, the number of possible derived conditions is almost unlimited due to the range of potential physical conditions that can be encountered in fluid flow problems. It is left to the producers of CFD software to document those conditions, when it is important that the user identifies the basic underlying type of the condition, *e.g.* fixed value or gradient.

## 4.1 Boundary mesh

Sec. 3.2 describes the computational mesh used by the finite volume method. It identifies a domain boundary by faces which are connected only to one cell. The boundary faces are grouped into *patches*, each under a unique name, corresponding to different regions of the boundary. Boundary conditions are then applied to each patch to provide a representative solution to the case of interest.

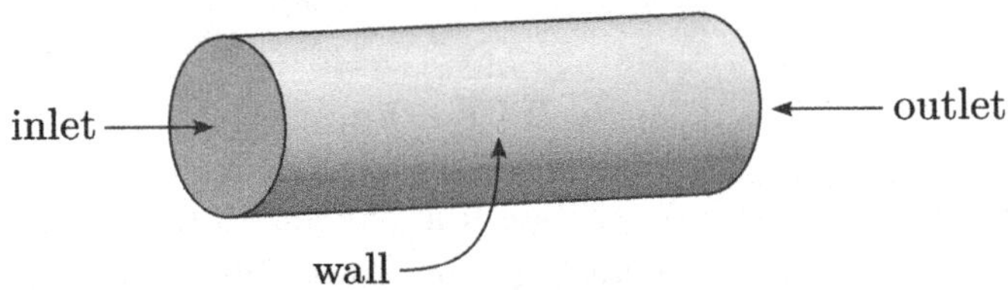


In the example of flow in a pipe, it would be logical to split the boundary into 3 patches in order to specify inflow, outflow and solid walls; and, *inlet*, *outlet* and *wall* would be logical names for these patches.

### Patch geometric data

The geometry of a patch is described using face data described in Sec. 3.3, including:

- the face area vector $\mathbf{S}_\mathrm{f}$, with area magnitude and direction $\mathbf{n}_\mathrm{f}$;
- the face unit normal vector $\mathbf{n}_\mathrm{f} = \mathbf{S}_\mathrm{f}/|\mathbf{S}_\mathrm{f}|$;
- the face centre $\mathbf{C}_\mathrm{f}$.

The cell connected to each patch face is denoted by the subscript "P", *e.g.* the cell centre is denoted $\mathbf{C}_\mathrm{P}$.

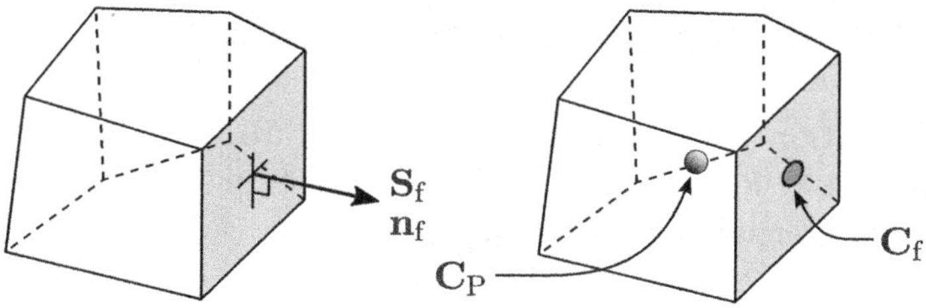


## Patch deltas

Sec. 3.8 describes the "delta" $\mathbf{\Delta d}$ for each face as the vector connecting the centres of its owner and neighbour cells. $\mathbf{\Delta d}$ is defined differently for a patch face since it has no neighbour cell.

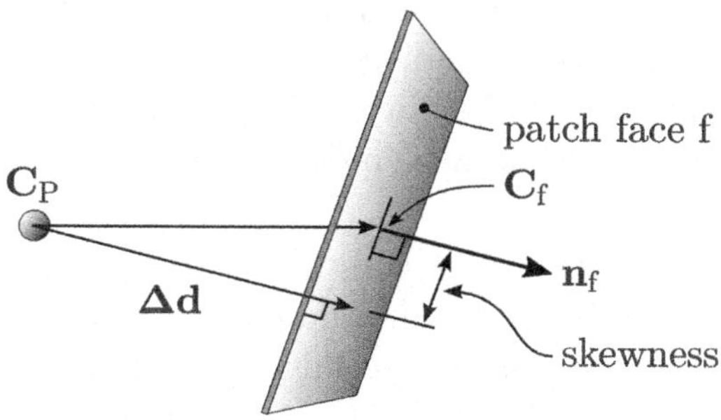


The delta is defined as the component of $\mathbf{C}_f - \mathbf{C}_P$ in the face normal direction. The surface gradient $\nabla_{\mathbf{n}} \mathbf{\Psi}_f$ is orthogonal to the face, which eliminates the error associated with non-orthogonality, at the expense of introducing a skewness error. Taking the inner product with $\mathbf{n}_f$ gives the magnitude which is then multiplied by $\mathbf{n}_f$ to assign the direction, *i.e.*

$$\mathbf{\Delta d} = \mathbf{n}_f(\mathbf{n}_f \bullet (\mathbf{C}_f - \mathbf{C}_P)) \tag{4.1}$$

The *delta coefficients* are $C_\Delta = 1/|\mathbf{\Delta d}|$, as defined in Sec. 3.8. The delta coefficient $C_\Delta$, representing "inverse distance", is a critical parameter in the discretisation of boundary conditions.

## 4.2 Fixed value and fixed gradient

In an equation for $\boldsymbol{\Psi}$ that is discretised to form a matrix equation $[\mathbf{A}][\boldsymbol{\Psi}] = [\mathbf{b}]$, there are two terms that include properties interpolated to faces:

- an advection term of the form $\nabla \bullet (\mathbf{u}\boldsymbol{\Psi})$ which is discretised by Eq. (3.8) in extensive form (see Sec. 3.5) as $\sum_{\mathrm{f}} \phi_{\mathrm{f}} \boldsymbol{\Psi}_{\mathrm{f}}$;
- a Laplacian term of the form $\nabla \bullet (\Gamma \nabla \boldsymbol{\Psi})$ which is discretised by Eq. (3.2) in extensive form as $\sum_{\mathrm{f}} |\mathbf{S}_{\mathrm{f}}| \, \Gamma_{\mathrm{f}} \nabla_{\mathbf{n}} \boldsymbol{\Psi}_{\mathrm{f}}$.

The advection term requires the value $\boldsymbol{\Psi}_{\mathrm{f}}$ and Laplacian term requires the surface normal gradient $\nabla_{\mathbf{n}} \boldsymbol{\Psi}_{\mathrm{f}}$ at faces. When a face is part of a boundary patch, any gradient $\nabla_{\mathbf{n}} \boldsymbol{\Psi}_{\mathrm{f}}$ and/or value $\boldsymbol{\Psi}_{\mathrm{f}}$ must be specified through boundary conditions.

The **fixed value**, or *Dirichlet* condition,† is the first type of boundary condition, where the boundary value $\boldsymbol{\Psi}_{\mathrm{b}}$ is specified. For example, at an inlet patch, we might specify a temperature $T_{\mathrm{b}} = 300$ K of the fluid flowing into the domain.

The **fixed gradient**, or *Neumann* condition,‡ is the second type, in which the gradient normal to the boundary $\nabla_{\mathbf{n}} \boldsymbol{\Psi}_{\mathrm{b}}$ is specified (where $\nabla_{\mathbf{n}} \equiv \mathbf{n} \bullet \nabla$). In many cases, the applied normal gradient is zero, which is a common condition applied to many fields, including $T$, at an outlet patch, *i.e.* where the fluid flows out of the domain.

---

†Peter Dirichlet, *Über einen neuen Ausdruck zur Bestimmung der Dictigkeit einerunendlich dünnen Kugelschale, wenn der Werth des Potentials derselben in jedem Punkte ihrer Oberfläche gegeben ist*, 1850.

‡after Carl Neumann 1832-1925.

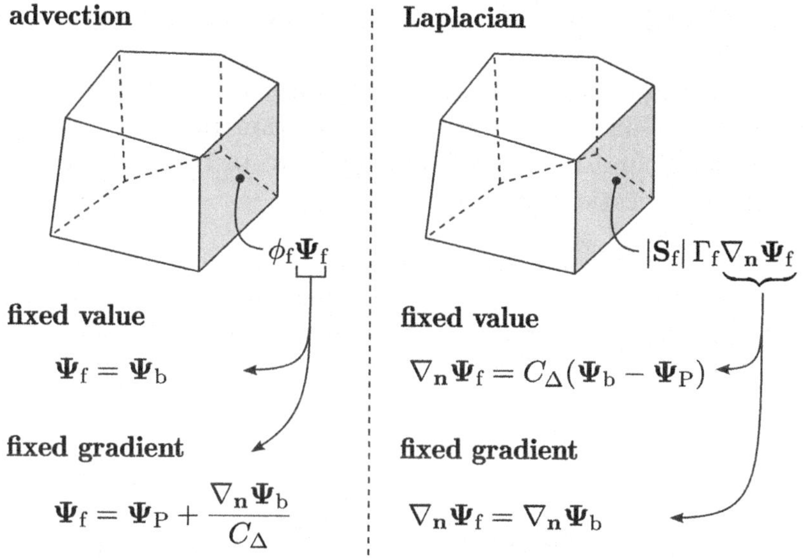


When an advection term is discretised, a fixed value condition is applied by substituting the face value $\mathbf{\Psi}_f$ with the patch value $\mathbf{\Psi}_b$, *i.e.* setting $\mathbf{\Psi}_f = \mathbf{\Psi}_b$. When a fixed gradient $\nabla_{\mathbf{n}} \mathbf{\Psi}_b$ is specified, the face value is expressed as follows, where $\mathbf{\Psi}_P$ is the value in the cell adjacent to each face:

$$\mathbf{\Psi}_f = \mathbf{\Psi}_P + \frac{\nabla_{\mathbf{n}} \mathbf{\Psi}_b}{C_\Delta}. \tag{4.2}$$

When a Laplacian term is discretised, a fixed gradient boundary condition is applied by substituting the face normal gradient $\nabla_{\mathbf{n}} \mathbf{\Psi}_f$ with the patch face normal gradient $\nabla_{\mathbf{n}} \mathbf{\Psi}_b$, *i.e.* setting $\nabla_{\mathbf{n}} \mathbf{\Psi}_f = \nabla_{\mathbf{n}} \mathbf{\Psi}_b$. When a fixed value $\mathbf{\Psi}_b$ is specified, the face normal gradient is expressed by

$$\nabla_{\mathbf{n}} \mathbf{\Psi}_f = C_\Delta (\mathbf{\Psi}_b - \mathbf{\Psi}_P). \tag{4.3}$$

## 4.3 Fundamentals of boundary conditions

The specification of boundary conditions is one of the most challenging tasks in setting up a CFD simulation. The range of possible boundary conditions is endless, to cover all of the potential applications and physics.

Setting boundary conditions is not an exact science but is guided by an basic specification using the fixed value (Dirichlet) and fixed gradient (Neumann) conditions introduced in Sec. 4.2.

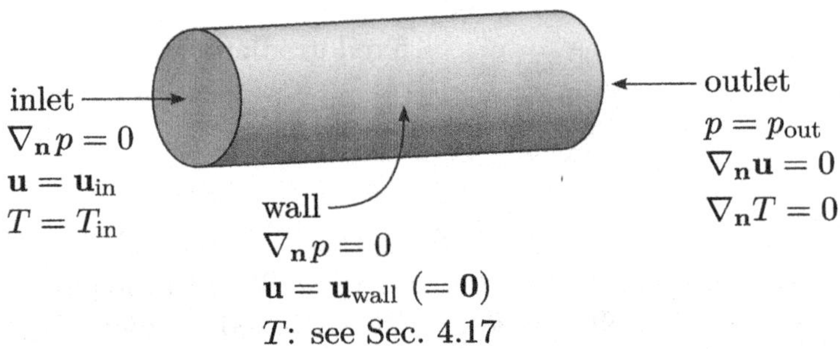


The figure above shows the basic specification for $p$, $\mathbf{u}$ and $T$ for incompressible subsonic flow, *e.g.* described by Eq. (2.47), Eq. (2.48) and Eq. (2.65).

The boundary conditions for $p$ and $\mathbf{u}$ require particular attention since they are coupled. The conditions on $T$ are independent of $p$ and $\mathbf{u}$ and are representative of other transported scalar fields, *e.g.* turbulent kinetic energy $k$.

The specification for inlets and outlets summarises as:

- zero gradient on $p$ at an inlet, fixed value on other variables;
- fixed value on $p$ at an outlet, zero gradient on other variables.

## Propagation of disturbances

A disturbance in a flow is simply any change from an equilibrium or steady solution. A disturbance at one location travels, or *propagates*, through the fluid.

The combination of boundary conditions at open boundaries, *i.e.* those excluding walls, relates to the propagation of disturbances. While disturbances are transported by advection with the flow, they propagate as waves at the speed of sound $c$.

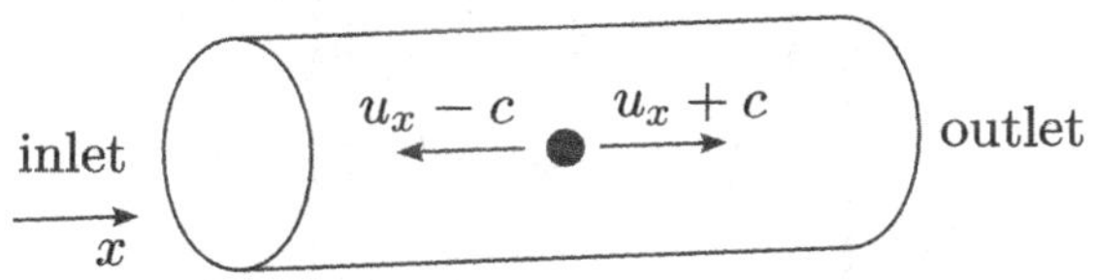


Sound waves can propagate disturbances against the direction of flow if it is *subsonic*, *i.e.* $u_x < c$. Disturbances must be able to propagate *outwards* through the inlet, which requires one variable not to be prescribed at the inlet. Similarly, they must be able propagate *inwards* from the outlet, which requires one variable to be prescribed at the outlet.

A pressure equation, combining mass and momentum conservation, describes wave propagation. For an incompressible fluid the wave speed $c = \infty$, since Eq. (2.48) contains no $\partial p / \partial t$ term. A disturbance at any point influences the solution everywhere in the domain instantaneously, as discussed in Sec. 2.22.

Pressure is then logically the variable on which to specify the boundary conditions to support wave propagation. Pressure is therefore: prescribed at the outlet, *i.e.* we specify fixed value condition; not prescribed at the inlet, *i.e.* we specify a fixed gradient condition, usually set to $\nabla_{\mathbf{n}} p = 0$.

## 4.4 Wall boundaries

The **no-slip** condition is generally applied at a solid wall which is impermeable (assuming $\nu \neq 0$). The condition is $\mathbf{u} = \mathbf{u}_{\text{wall}}$, where $\mathbf{u}_{\text{wall}}$ is the velocity of the wall, which is usually stationary with $\mathbf{u}_{\text{wall}} = \mathbf{0}$. The proof behind the no-slip condition is that it predicts a pressure drop along tubes of small diameter which matches experiments.†

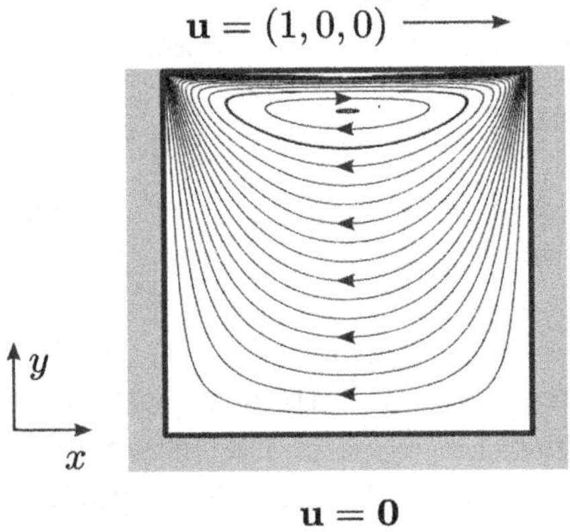


The 2D, lid-driven cavity is a flow problem in which no-slip conditions are applied at all boundaries. It provides insight into the boundary condition for $p$ at a wall. From Eq. (2.47) for an incompressible fluid, with $\nu$ = constant, the component normal to the domain boundary is

$$\nabla_{\mathbf{n}} p = \mathbf{n} \bullet \left[ \mathbf{b} + \nabla \bullet (\nu \nabla \mathbf{u}) - \nabla \bullet (\mathbf{u}\mathbf{u}) - \frac{\partial \mathbf{u}}{\partial t} \right]. \tag{4.4}$$

At the solid wall boundary, the last 2 terms in Eq. (4.4) disappear since $\mathbf{n} \bullet \mathbf{u} = 0$, reducing the gradient condition to

$$\nabla_{\mathbf{n}} p = \mathbf{n} \bullet \mathbf{b} + \mathbf{n} \bullet \nabla \bullet (\nu \nabla \mathbf{u}). \tag{4.5}$$

†Jean Poiseuille, *Recherches expérimentales sur le mouvement des liquides dans les tubes de très petits diamètres I-IV*, 1840.

The second term can be written ($\nu$ = constant) $\nu\,\partial^2 u_n/\partial n^2$, so is only non-zero where there is flow normal to the boundary in its vicinity, *e.g.* at the corners of the cavity in our example. The term is usually small and its calculation involves extrapolation from the internal solution which often causes instability, so it is generally ignored.

A body force **b**, *e.g.* gravity, is generally prescribed so it does not introduce instability. Where it is significant, it must be included in the boundary condition, *i.e.* $\nabla_{\mathbf{n}} p = \mathbf{n} \bullet \mathbf{b}$.

Otherwise, in the absence of a body force, we reach the standard form of boundary condition for pressure at a wall, $\nabla_{\mathbf{n}} p = 0$.

## Fixing pressure

With only fixed *gradient* conditions on pressure at the boundary, the pressure *value* is not fixed at any point in solution domain. The solution is not unique, as shown in the 1D example below with gradient conditions at both ends.

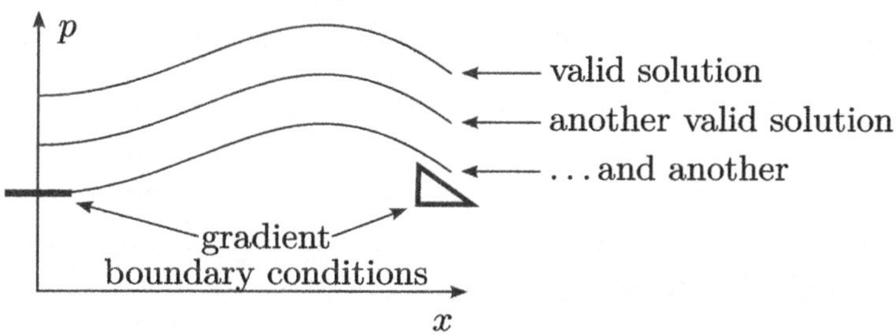


To achieve a unique solution, $p$ must then be fixed to a reference value $p_0$ at a reference cell $r$ in the domain. To achieve this, the diagonal coefficient $a_{r,r}$ is doubled and $a_{r,r}p_0$ is added to source $b_r$, in the matrix equation $[\,\mathbf{A}\,][\,p\,] = [\,\mathbf{b}\,]$ described by Eq. (3.1). This minimal change "pins" the solution to $p = p_0$ in cell $r$.

## 4.5 Inlets and outlets

A basic set of boundary conditions was introduced in Sec. 4.3 for incompressible subsonic flow.

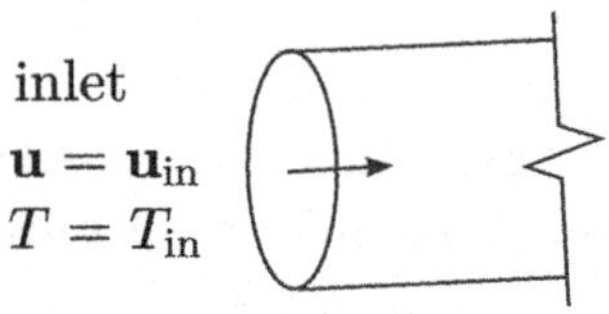


At an *inlet*, fields are generally specified as fixed value. From a "physical" perspective, this is justified by the fact that disturbances propagate in the direction of flow so must be specified at the upstream boundary.

Advection requires interpolation of variables from cell centres ($\mathbf{\Psi}$) to faces ($\mathbf{\Psi}_{\mathrm{f}}$). Some degree of upwind interpolation, *e.g.* as part of a limited scheme, is generally required. At a face at an inlet patch there is no upwind cell, so values $\mathbf{\Psi}_{\mathrm{f}}$ must be specified at inlet faces instead. From a "numerical" perspective, this justifies the fixed value boundary condition at an inlet.

For incompressible, subsonic flow, $p$ is the exception. A fixed gradient condition allows disturbances to propagate upstream through the inlet as sound waves. The gradient condition is further justified since there is no advection of $p$, see Eq. (2.48), so no upwind interpolation is required at the inlet patch faces.

At an *outlet*, the converse is then true: fields are specified as fixed gradient condition, with the exception of $p$ which is fixed value. The outlet conditions ultimately dictate the traction force combining Eq. (2.16), Eq. (2.33), Eq. (2.46) and Eq. (2.41) as follows:

$$\mathbf{t} = \mu \left( \nabla_{\mathbf{n}} \mathbf{u} + \nabla u_n \right) - p\mathbf{n} \,. \tag{4.6}$$

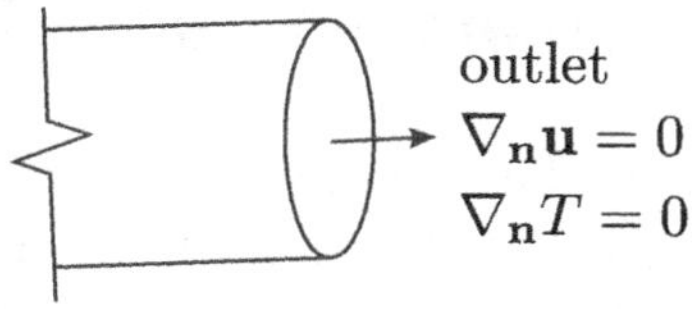


The standard condition applied to $\mathbf{u}$ is $\nabla_{\mathbf{n}}\mathbf{u} = 0$. By Eq. (4.6), this results in a uniform *normal* traction force corresponding to the outlet pressure $p$. The traction force *tangential* to the outlet $= \mu \partial u_n / \partial m$, where $m$ is the tangential direction and $u_n$ is the normal component of $\mathbf{u}$.

## Supersonic conditions

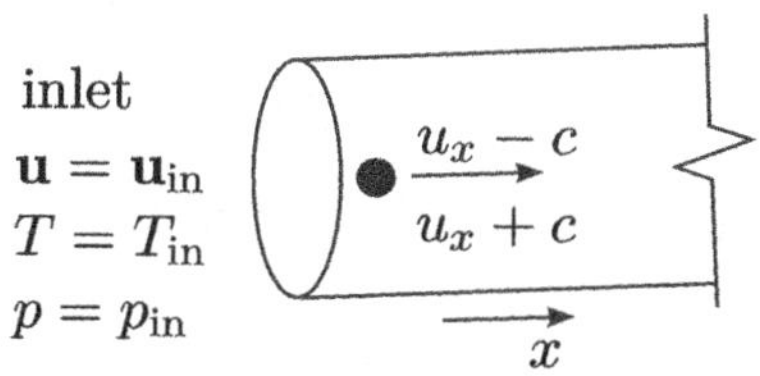


If the fluid is compressible and the flow speed is *supersonic* at the inlet, *i.e.* $u_x > c$, waves can no longer propagate outwards through the inlet boundary. When this occurs, a fixed value $p$ must be specified at the inlet.

Similarly if the flow is supersonic at an outlet, all disturbances propagate through the outlet. In that case, $p$ cannot be specified, *i.e.* a condition on the normal gradient $\nabla_{\mathbf{n}}p$ is applied.

Some flow domains combine subsonic flow at the inlet and supersonic flow at the outlet. When this occurs, a gradient condition is required for $p$ at both boundaries, leaving $p$ under-specified. This problem is best overcome by moving the outlet boundary sufficiently downstream for the flow to expand to subsonic speed. This allows $p$ to be specified through a fixed value condition.

## 4.6 Free (entrainment) boundaries

The basic set of boundary conditions for subsonic incompressible flow, introduced in Sec. 4.3, relates to inlet, outlet and wall boundary regions. Sometimes a single boundary patch occupies a location which experiences both inflow and outflow.

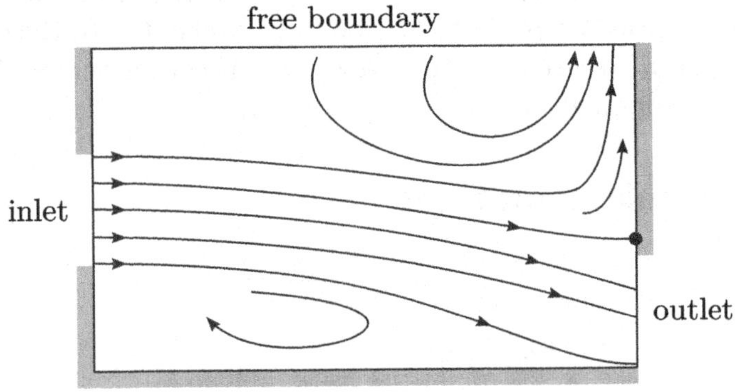


The figure above shows a 2D domain with an inlet and outlet, left and right respectively. The top boundary is *free* to allow:

- inflow from the centre to the left side of the patch, through entrainment of fluid by the main flow from the inlet;
- outflow at the right side of the patch, driven by a pressure gradient that emanates from the point where the flow impinges on the right wall.

Boundary conditions for $p$ and $\mathbf{u}$ are therefore required which support both inflow and outflow at a single patch. The inlet flow speed is determined from the flow solution rather than being prescribed at the patch. This suggests a gradient condition on $\mathbf{u}$, which is generally prescribed at an outlet.

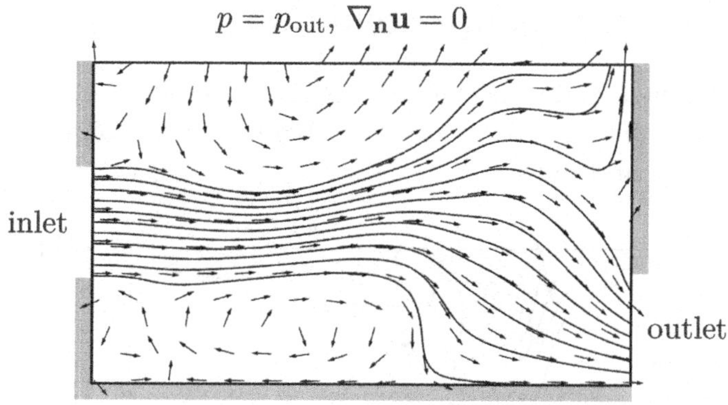


At a free/entrainment boundary, the basic outflow conditions, $p = p_{\text{out}}$ and $\nabla_{\mathbf{n}}\mathbf{u} = 0$, are generally **not** recommended. The figure above shows a snapshot from a simulation using those conditions. The solution oscillates between changing levels of inflow and outflow over the open boundary; the figure below shows the flow leading up to the snapshot above.

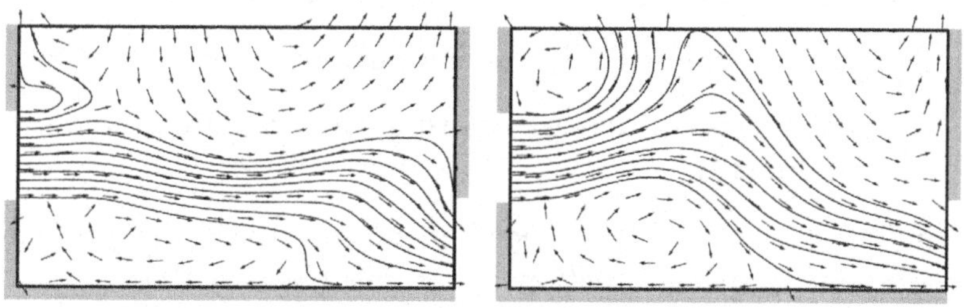

The is no convergence towards a stable solution in this example of a simple flow in a regular 2D geometry. Moreover, solutions tend to diverge and "blow up" if the basic outflow conditions are applied at a free boundary for geometries and meshes that are even moderately complex. In the following sections we look at some alternative conditions that are applied at a free boundary to deliver stable, robust solutions.

## 4.7 Total pressure condition

Sec. 4.6 concluded that the basic outflow conditions — $p = p_{\text{out}}$ and $\nabla_{\mathbf{n}}\mathbf{u} = 0$ — are generally unstable for a free boundary with both inflow and outflow.

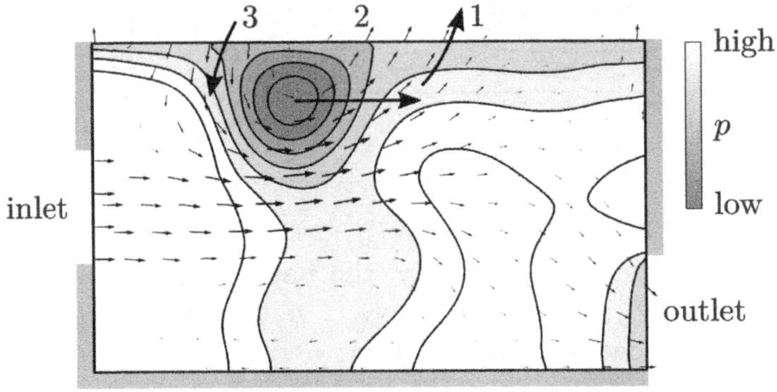


The conditions are unstable due to pressure fluctuations at the boundary, shown above. Flows oscillates in and out, shown by 3 stages, creating a vortex that travels from left to right:

1. the pressure gradient, decreasing outward, causes outflow;
2. the outflow speed increases causing the pressure gradient to change direction;
3. inflow begins and the speed increases, until the pressure gradient changes direction, returning back to step 1.

The **total pressure** boundary condition improves the stability of solutions. It is a fixed value type, calculated according to:

$$p = \begin{cases} p_0 & \text{for outflow;} \\ p_0 - |\mathbf{u}^2|/2 & \text{for inflow.} \end{cases} \tag{4.7}$$

The specified total pressure, $p_0$, can be imagined as the fluid pressure under quiescent conditions far from the free boundary, which decreases as the fluid accelerates towards the boundary. Note that Eq. (4.7) is written for the incompressible assumption, Eq. (2.47), where $p$ and $p_0$ are kinematic, *i.e.* divided by $\rho$.

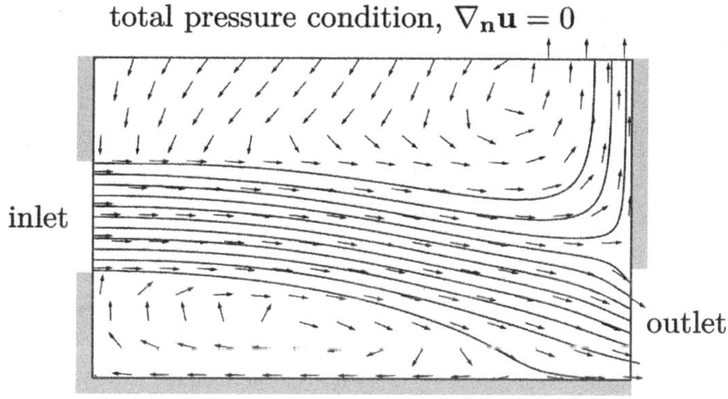


The solution using the total pressure condition converges to the flow field shown above. The critical effect of this boundary condition is that, the boundary $p$ decreases by $|\mathbf{u}^2|/2$ as the inflow speed $|\mathbf{u}|$ increases. This reduces the pressure gradient driving inflow, which moderates the increase in inflow speed, enabling it to settle to a stable level.

## Total pressure for high speed flow

The total pressure condition can be applied to high-speed flow of a compressible gas. The calculation of $p$ for inflow is simply replaced with the 1D isentropic flow equation,†

$$p = p_0 \left[1 + \frac{\gamma - 1}{2}\mathrm{Ma}^2\right]^{\gamma/(1-\gamma)}, \tag{4.8}$$

where Mach number is $\mathrm{Ma} = |\mathbf{u}|/c$ and $\gamma = (R + c_V)/c_V$.

†Ascher Shapiro, *The dynamics and thermodynamics of compressible fluid flow, Vol. 1, Ch. 4*, 1953.

## 4.8 Numerical framework

The fixed value and fixed gradient conditions described in Sec. 4.2 can be combined to form a general numerical framework for boundary conditions.

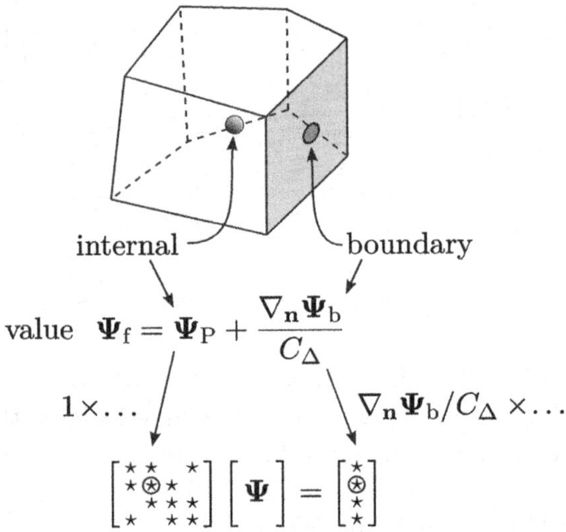


The contributions from the boundary conditions to the matrix equation $[\mathbf{A}][\Psi] = [\mathbf{b}]$, by discretisation of advection and Laplacian terms, can be generalised as:

- "*internal*" contributions to $[\mathbf{A}]$, from terms including the cell value $\Psi_{\mathrm{P}}$;
- "*boundary*" contributions to $[\mathbf{b}]$, from terms without $\Psi_{\mathrm{P}}$.

The example above shows Eq. (4.2) for the face value, required at a boundary for advection discretisation, in the case of a *fixed gradient* boundary condition. The internal "factor" on $\Psi_{\mathrm{P}}$ is 1, which is multiplied by $\phi_{\mathrm{f}}$ for the contribution to the respective diagonal coefficient in $[\mathbf{A}]$, as in the example in Sec. 3.24.

The boundary factor is $\nabla_{\mathbf{n}}\mathbf{\Psi}_{\mathrm{b}}/C_{\Delta}$, which is similarly multiplied by $\phi_{\mathrm{f}}$ for the contribution to $[\,\mathbf{b}\,]$.

For the *fixed value* condition $\mathbf{\Psi}_{\mathrm{f}} = \mathbf{\Psi}_{\mathrm{b}}$ with *advection*, the boundary factor is $\mathbf{\Psi}_{\mathrm{b}}$ and an internal factor is 0.

Laplacian discretisation requires the surface normal gradient on the faces. A *fixed gradient* condition delivers an equivalent boundary factor of $\nabla_{\mathbf{n}}\mathbf{\Psi}_{\mathrm{b}}$ to $[\,\mathbf{b}\,]$ and an internal factor of 0.

$$\nabla_{\mathbf{n}}\mathbf{\Psi}_{\mathrm{f}} = C_{\Delta}(\mathbf{\Psi}_{\mathrm{b}} - \mathbf{\Psi}_{\mathrm{P}})$$

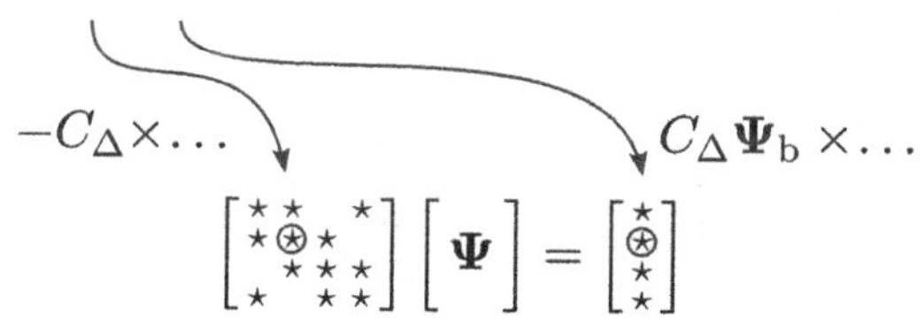


With a *fixed value* $\mathbf{\Psi}_{\mathrm{b}}$, the face normal gradient Eq. (4.3) gives an internal factor of $-C_{\Delta}$ and a boundary factor of $C_{\Delta}\mathbf{\Psi}_{\mathrm{b}}$. Both are multiplied by $|\mathbf{S}_{\mathrm{f}}|\Gamma_{\mathrm{f}}$ in their contributions to $[\,\mathbf{b}\,]$ and the diagonal coefficient of $\mathbf{\Psi}_{\mathrm{P}}$ in $[\,\mathbf{A}\,]$, as shown in the Laplacian discretisation in Sec. 3.24.

The table below summarises: the "value" internal and boundary factors, contributing to the respective matrix coefficients with advection discretisation; and equivalent "gradient" factors relating to Laplacian discretisation. This provides a framework which can be extended to more complex conditions.

| term | factor | fixed value | fixed gradient |
|---|---|---|---|
| advection | value internal | 0 | 1 |
| advection | value boundary | $\mathbf{\Psi}_{\mathrm{b}}$ | $\nabla_{\mathbf{n}}\mathbf{\Psi}_{\mathrm{b}}/C_{\Delta}$ |
| Laplacian | gradient internal | $-C_{\Delta}$ | 0 |
| Laplacian | gradient boundary | $C_{\Delta}\mathbf{\Psi}_{\mathrm{b}}$ | $\nabla_{\mathbf{n}}\mathbf{\Psi}_{\mathrm{b}}$ |

## 4.9 Mixed fixed value/gradient

Sec. 4.8 concluded with a table of factors for the contributions to coefficients of $[\,\mathbf{A}\,]$ and $[\mathbf{b}]$ for the fixed value and fixed gradient boundary conditions.

It distinguishes between contributions for the discretisation of an advection term which requires *values* at faces, and the Laplacian term which requires the normal *gradient*.

A **mixed** fixed value/gradient condition is defined by introducing a *value fraction* $v$ for which

$$v = \begin{cases} 0 & \text{represents fixed gradient,} \\ 1 & \text{represents fixed value.} \end{cases} \tag{4.9}$$

Within the range $0 < v < 1$ the condition operates in between fixed value and fixed gradient. The mixed condition is simply implemented by blending the fixed value and fixed gradient contributions by $v$, as shown in the table below.

| factor | mixed |
|---|---|
| value internal | $(1 - v)$ |
| value boundary | $v\boldsymbol{\Psi}_\mathrm{b} + (1 - v)\nabla_\mathbf{n}\boldsymbol{\Psi}_\mathrm{b}/C_\Delta$ |
| gradient internal | $-vC_\Delta$ |
| gradient boundary | $vC_\Delta\boldsymbol{\Psi}_\mathrm{b} + (1 - v)\nabla_\mathbf{n}\boldsymbol{\Psi}_\mathrm{b}$ |

This mixed condition provides the framework for a boundary condition that can switch between a fixed value $\boldsymbol{\Psi}_\mathrm{b}$ and fixed gradient $\nabla_\mathbf{n}\boldsymbol{\Psi}_\mathrm{b}$, by changing $v$. Switching is often based on flow direction, $v = 1$ corresponding to inflow and $v = 0$ to outflow.

Some boundary conditions can operate in the range of value fractions $0 < v < 1$. The Robin condition, described next, can also be expressed as a mixed condition with varying $v$.

## Robin condition

The *Robin* condition[†] combines the value and normal gradient at the boundary through an expression:

$$\boldsymbol{\Psi} + a\nabla_{\mathbf{n}}\boldsymbol{\Psi} = \boldsymbol{\Psi}_0, \tag{4.10}$$

where $a$ is a scalar coefficient with units of length and $\boldsymbol{\Psi}_0$ is some constant value of $\boldsymbol{\Psi}$.

The Robin condition can be treated like the mixed condition by relating $\boldsymbol{\Psi}_0$ to a reference fixed value $\boldsymbol{\Psi}_\mathrm{b}$ and gradient $\nabla_{\mathbf{n}}\boldsymbol{\Psi}_\mathrm{b}$ at a boundary, according to $\boldsymbol{\Psi}_0 = \boldsymbol{\Psi}_\mathrm{b} + a\nabla_{\mathbf{n}}\boldsymbol{\Psi}_\mathrm{b}$.

Substituting for $\boldsymbol{\Psi}_0$ in Eq. (4.10) and making $\boldsymbol{\Psi}$ the subject of the equation gives:

$$\boldsymbol{\Psi} = \frac{1}{1 + aC_\Delta}\boldsymbol{\Psi}_\mathrm{b} + \frac{a}{1 + aC_\Delta}\nabla_{\mathbf{n}}\boldsymbol{\Psi}_\mathrm{b} + \frac{aC_\Delta}{1 + aC_\Delta}\boldsymbol{\Psi}_\mathrm{P}$$

Comparison with the value factors in the previous table shows that the Robin condition corresponds to the mixed condition with the value fraction $v = 1/(1 + aC_\Delta)$.

In this form, $\boldsymbol{\Psi}_\mathrm{b}$ and $\nabla_{\mathbf{n}}\boldsymbol{\Psi}_\mathrm{b}$ relate to the limits $v \to 1$ and $v \to 0$, respectively. Values can be selected in these limits to represent the physics of the condition.

In many cases the reference gradient is $\nabla_{\mathbf{n}}\boldsymbol{\Psi}_\mathrm{b} = 0$ such that $\boldsymbol{\Psi}_0 = \boldsymbol{\Psi}_\mathrm{b}$ in Eq. (4.10). For example a condition for temperature $T + a\nabla_{\mathbf{n}}T = T_\mathrm{b}$ would tend to $T = T_\mathrm{b}$ as $a \to 0$ and $\nabla_{\mathbf{n}}T = 0$ as $a \to \infty$.

The value fraction $v$ includes $C_\Delta$, so the condition operates "in the middle" between the fixed value and gradient when $a$ is the same order of magnitude as the boundary cell height.

[†]after Victor Robin (1855-1897).

## 4.10 Mixed inlet-outlet condition

The **inlet-outlet** boundary condition is the most basic example of the mixed fixed value/gradient type, described in Sec. 4.9. The condition sets the reference gradient $\nabla_{\mathbf{n}}\mathbf{\Psi}_{\mathrm{b}} = 0$ and uses a specified reference value $\mathbf{\Psi}_{\mathrm{b}}$. The value fraction is then set to

$$v = \begin{cases} 0 & \text{for outflow,} \\ 1 & \text{for inflow.} \end{cases} \tag{4.11}$$

The flow direction is established from the sign of the volumetric or mass flux $\phi_{\mathrm{f}}$ at each boundary face, described in Sec. 2.8, by

$$\begin{aligned} \phi_{\mathrm{f}} &\geq 0 \quad \text{for outflow,} \\ \phi_{\mathrm{f}} &< 0 \quad \text{for inflow.} \end{aligned} \tag{4.12}$$

The inlet-outlet condition is generally very useful for *scalar* fields, *e.g.* turbulence fields, $T$, *etc.* It has an immediate practical use at a free boundary, *e.g.* in the case introduced in Sec. 4.6.

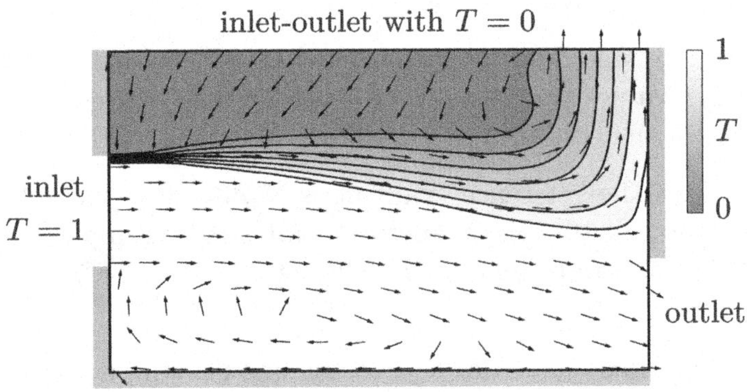


The figure shows the solution of Eq. (2.65), converged over time with $r = 0$ and unity Prandtl number Pr, see Sec. 2.21. The fixed condition $T = 1$ is applied at the inlet and a zero gradient condition $\nabla_{\mathbf{n}} T = 0$ at the walls.

At the free boundary, the inlet-outlet condition enables $T$ to be specified where inflow occurs. The inlet value in the example is set to $T = 0$; the image shows mixing of fluids at different temperatures, from the inlet and entrained at the free boundary.

## Numerical benefit of inlet-outlet

Boundaries may be described "inlet" and "outlet" based on the expectation of the flow direction during a simulation. But the flow direction may not always happen as expected.

In the case of an outlet, for example, inflow might occur during a simulation. For example, at the start of a simulation, the initial conditions may induce inflow before the internal flow is established. Localised inflow can also occur when rotating flow structures pass through an outlet boundary, *e.g.* when a bluff body sheds vortices, as shown below.

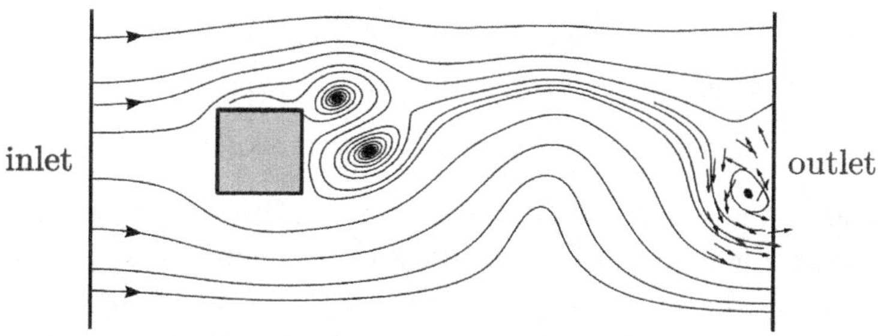


Where inflow occurs, the inlet-outlet condition can switch to the fixed value type to ensure stability, as discussed in Sec. 4.5. The inlet-outlet condition is therefore commonly applied to scalar fields (except $p$), at a boundary which is notionally an outlet, to avoid numerical instability associated with unexpected inflow.

## 4.11 Transform condition

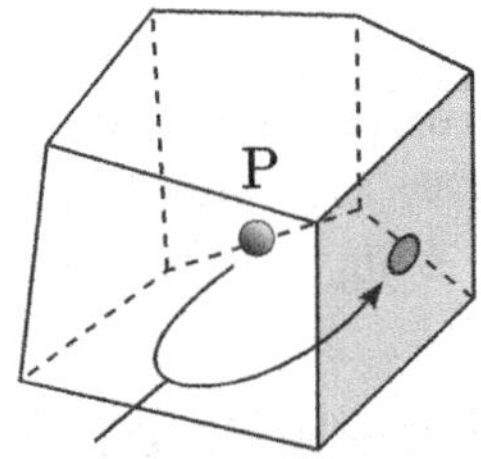


transformation: $\sum_{i=1}^{N} G(\mathbf{T}_i, \mathbf{\Psi}_\mathrm{P})$

Some boundary conditions represent $\mathbf{\Psi}_\mathrm{f}$ at the boundary as a transformation of the cell value $\mathbf{\Psi}_\mathrm{P}$. They can be expressed in terms of a general **transform** condition

$$\mathbf{\Psi}_\mathrm{f} = \sum_{i=1}^{N} G(\mathbf{T}_i, \mathbf{\Psi}_\mathrm{P}), \tag{4.13}$$

where $G(\mathbf{T}, \mathbf{\Psi}_\mathrm{P})$ is the geometric transformation of variable $\mathbf{\Psi}_\mathrm{P}$ by a tensor $\mathbf{T}$. The transformation is calculated as follows:

$$G(\mathbf{T}, \mathbf{\Psi}_\mathrm{P}) = \begin{cases} \mathbf{\Psi}_\mathrm{P} & \text{for scalar } \mathbf{\Psi}; \\ \mathbf{T} \bullet \mathbf{\Psi}_\mathrm{P} & \text{for vector } \mathbf{\Psi}; \\ \mathbf{T} \bullet \mathbf{\Psi}_\mathrm{P} \bullet \mathbf{T}^\mathrm{T} & \text{for tensor } \mathbf{\Psi}. \end{cases} \tag{4.14}$$

When $\mathbf{\Psi}$ is a scalar, the transform condition is equivalent to a zero gradient condition. Otherwise, it is implemented so that terms in $\mathbf{\Psi}_\mathrm{P}$ in Eq. (4.13) contribute to coefficients in $[\,\mathbf{A}\,]$. These contributions are implicit which improves convergence when solving the matrix equation $[\,\mathbf{A}\,][\,\mathbf{\Psi}\,] = [\mathbf{b}]$.

A factor $\mathbf{\Lambda}$ is introduced to specify the contribution to internal coefficients. It represents a single internal coefficient for each component of $\mathbf{\Psi}$ so has the same rank as $\mathbf{\Psi}$, *i.e.* it is a vector when $\mathbf{\Psi}$ is a vector, and a tensor when $\mathbf{\Psi}$ is a tensor.

The multiplication of each coefficient of $\mathbf{\Lambda}$ by its respective component of $\mathbf{\Psi}_\mathrm{P}$ is denoted by by $\mathbf{\Lambda} \times \mathbf{\Psi}_\mathrm{P}$. In the case of vector $\mathbf{\Psi}_\mathrm{P}$, this "component multiplication" is

$$\mathbf{\Lambda} \times \mathbf{\Psi}_\mathrm{P} = (\mathbf{\Lambda}_x \mathbf{\Psi}_{\mathrm{P}x}, \mathbf{\Lambda}_y \mathbf{\Psi}_{\mathrm{P}y}, \mathbf{\Lambda}_z \mathbf{\Psi}_{\mathrm{P}z}).$$

For advection discretisation, the face value is represented as

$$\mathbf{\Psi}_\mathrm{f} = \underbrace{\mathbf{\Psi}_\mathrm{e} - (\mathbf{1} - \mathbf{\Lambda}) \times \mathbf{\Psi}_\mathrm{P}}_{\text{boundary, explicit}} + \underbrace{(\mathbf{1} - \mathbf{\Lambda})}_{\text{internal}} \times \mathbf{\Psi}_\mathrm{P}, \tag{4.15}$$

where $\mathbf{\Psi}_\mathrm{e}$ is an explicit boundary value, *calculated* from the expression in Eq. (4.13) using the current values $\mathbf{\Psi}_\mathrm{P}$. The other explicit term uses the current $\mathbf{\Psi}_\mathrm{P}$ with "$\mathbf{1}$" denoting $(1, 1, 1)$ for a vector $\mathbf{\Psi}$ and $\mathbf{I}$ for a tensor.

Laplacian discretisation requires the face normal gradient $\nabla_\mathbf{n} \mathbf{\Psi}_\mathrm{f}$. Combining $\nabla_\mathbf{n} \mathbf{\Psi}_\mathrm{f} = C_\Delta(\mathbf{\Psi}_\mathrm{f} - \mathbf{\Psi}_\mathrm{P})$ with Eq. (4.15) gives

$$\nabla_\mathbf{n} \mathbf{\Psi}_\mathrm{f} = \underbrace{(\nabla_\mathbf{n} \mathbf{\Psi})_\mathrm{e} + C_\Delta \mathbf{\Lambda} \times \mathbf{\Psi}_\mathrm{P}}_{\text{boundary, explicit}} - \underbrace{C_\Delta \mathbf{\Lambda}}_{\text{internal}} \times \mathbf{\Psi}_\mathrm{P}, \tag{4.16}$$

where the explicit gradient is calculated by

$$(\nabla_\mathbf{n} \mathbf{\Psi})_\mathrm{e} = C_\Delta(\mathbf{\Psi}_\mathrm{e} - \mathbf{\Psi}_\mathrm{P}). \tag{4.17}$$

The transform condition is summarised within the table below by the value and gradient contributions.

| factor | transform |
|---|---|
| value internal | $\mathbf{1} - \mathbf{\Lambda}$ |
| value boundary | $\mathbf{\Psi}_\mathrm{e} - (\mathbf{1} - \mathbf{\Lambda}) \times \mathbf{\Psi}_\mathrm{P}$ |
| gradient internal | $-C_\Delta \mathbf{\Lambda}$ |
| gradient boundary | $(\nabla_\mathbf{n} \mathbf{\Psi})_\mathrm{e} + C_\Delta \mathbf{\Lambda} \times \mathbf{\Psi}_\mathrm{P}$ |

## 4.12 Symmetry condition

The transform boundary condition was presented in Sec. 4.11. It provides a convenient framework for implementing boundary conditions that represent a *geometric constraint*, including the *symmetry* condition.

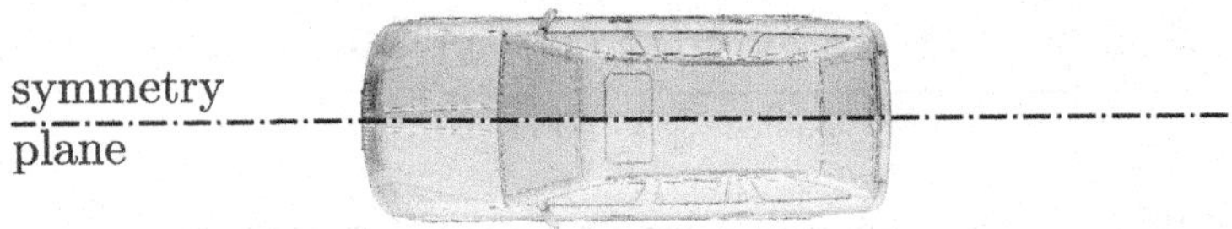


The **symmetry** condition is suitable for simulations where the geometry contains a plane of symmetry and the flow field is assumed symmetric. By generating a mesh on one side of a plane of symmetry and applying the symmetry condition, the number of cells, and hence solution time, is reduced.

In the context of a wall boundary, the symmetry condition is also equivalent to **slip** (as opposed to the common no-slip condition).

A symmetry plane is a transform condition so when the solution variable $\boldsymbol{\Psi}$ is a scalar, it reduces to zero gradient. For a vector, *e.g.* $\mathbf{u}$, the condition is zero gradient tangential to the plane, and zero fixed value normal to the plane.

When $\boldsymbol{\Psi}$ is a tensor, the boundary condition requires a more precise definition, which can also be applied to a vector $\boldsymbol{\Psi}$. The boundary $\boldsymbol{\Psi}_{\mathrm{f}}$ can be considered as the mean of the adjacent cell $\boldsymbol{\Psi}_{\mathrm{P}}$ and the mirror image $\boldsymbol{\Psi}_{\mathrm{P}}$ transformed by the reflective transformation tensor $\mathbf{I} - 2\mathbf{n}_{\mathrm{f}}\mathbf{n}_{\mathrm{f}}$, *i.e.*

$$\boldsymbol{\Psi}_{\mathrm{f}} = \frac{1}{2}\left[\boldsymbol{\Psi}_{\mathrm{P}} + G(\mathbf{I} - 2\mathbf{n}_{\mathrm{f}}\mathbf{n}_{\mathrm{f}}, \boldsymbol{\Psi}_{\mathrm{P}})\right]. \tag{4.18}$$

Here, $\mathbf{n}_{\mathrm{f}}$ is the unit normal vector on the boundary face.

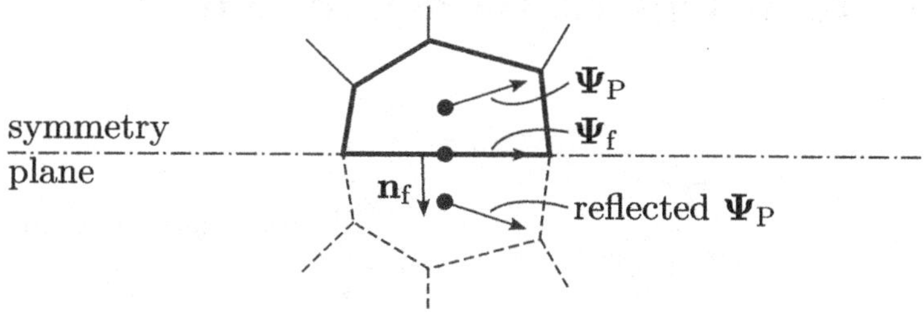


Using the notation in Sec. 4.11, the explicit boundary value $\mathbf{\Psi}_\mathrm{e}$ is calculated using current $\mathbf{\Psi}_\mathrm{P}$ from Eq. (4.18) and the explicit gradient $(\nabla_\mathbf{n}\mathbf{\Psi})_\mathrm{e}$ is calculated by Eq. (4.17).

Comparing Eq. (4.18) with the transform condition of Sec. 4.11, the factor $\mathbf{\Lambda}$ corresponds to the tensor $\mathbf{n}_\mathrm{f}\mathbf{n}_\mathrm{f}$. For a vector field, a factor which gives good solution convergence is

$$\mathbf{\Lambda} = \mathrm{diag}(\sqrt{\mathbf{n}_\mathrm{f}\mathbf{n}_\mathrm{f}}) = (|\mathbf{n}_{\mathrm{f}x}|, |\mathbf{n}_{\mathrm{f}y}|, |\mathbf{n}_{\mathrm{f}z}|), \tag{4.19}$$

where "diag(**T**)" is the vector of the diagonal components of tensor **T**.

For a tensor field, good convergence is achieved with a tensor $\mathbf{\Lambda}$ calculated as the outer product of $\mathbf{\Lambda}$ for a vector, *i.e.* denoting the vector factor by $\mathbf{\Lambda}_\mathrm{vec}$, the tensor factor is $\mathbf{\Lambda} = \mathbf{\Lambda}_\mathrm{vec}\mathbf{\Lambda}_\mathrm{vec}$.

## Orthogonality condition

The axes $Ox$, $Oy$, $Oz$, introduced in Sec. 2.1, must remain orthogonal under a transformation. It requires the transpose of the transformation tensor **T** to equal its inverse, *i.e.* $\mathbf{T}^\mathrm{T} = \mathbf{T}^{-1}$.

The orthogonality condition is therefore $\mathbf{T} \bullet \mathbf{T}^\mathrm{T} = \mathbf{I}$. The reflective transformation $\mathbf{T} = \mathbf{I} - 2\mathbf{n}_\mathrm{f}\mathbf{n}_\mathrm{f}$ satisfies the orthogonality condition since

$$\begin{aligned}\mathbf{T} \bullet \mathbf{T}^\mathrm{T} &= (\mathbf{I} - 2\mathbf{n}_\mathrm{f}\mathbf{n}_\mathrm{f}) \bullet (\mathbf{I} - 2\mathbf{n}_\mathrm{f}\mathbf{n}_\mathrm{f})\\ &= \mathbf{I} - 4\mathbf{n}_\mathrm{f}\mathbf{n}_\mathrm{f} + 4\mathbf{n}_\mathrm{f}\mathbf{n}_\mathrm{f} \bullet \mathbf{n}_\mathrm{f}\mathbf{n}_\mathrm{f} = \mathbf{I}.\end{aligned}$$

## 4.13 Axisymmetric (wedge) condition

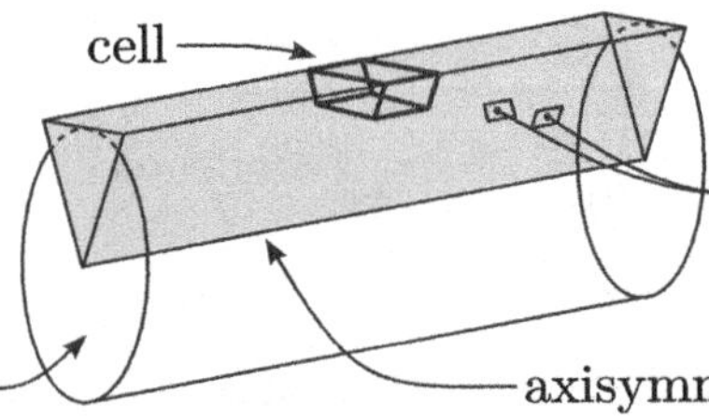


There are some fluid flow problems for which the geometry is axisymmetric. Assuming the flow solution is axisymmetric, *i.e.* fields do not change in the circumferential direction, the computational mesh can be formed of a wedge-shaped slice of the flow geometry.

This type of mesh for axisymmetric solution contains one cell across the circumferential direction, which reduces the number of cells to two dimensions in the axial and radial directions.

This approach to axisymmetric solution introduces a geometric error due the faces normal to the radial direction being flat. This error reduces with decreasing wedge angle; in practice, the error can be considered negligible for an angle of 1°.

The **wedge** boundary condition is applied to the two sloping side patches. It transforms cell values $\mathbf{\Psi}_{\mathrm{P}}$ to the patch faces using a rotational transformation tensor $\mathbf{R}$ by

$$\mathbf{\Psi}_{\mathrm{f}} = G(\mathbf{R}_{\mathrm{f}}, \mathbf{\Psi}_{\mathrm{P}}). \tag{4.20}$$

$\mathbf{R}_{\mathrm{f}}$ defines a rotation between the unit vector $\mathbf{n}$ in the circumferential direction at the cell centre and the unit face normal vector $\mathbf{n}_{\mathrm{f}}$ by $\mathbf{n}_{\mathrm{f}} = \mathbf{R}_{\mathrm{f}} \bullet \mathbf{n}$.

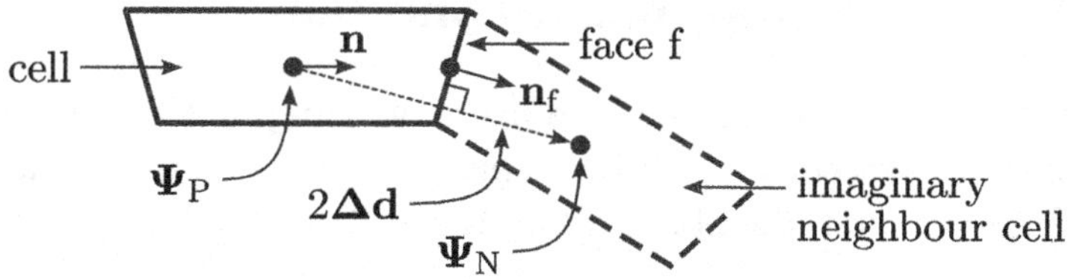


The wedge condition uses the general transform framework of Sec. 4.11, with the explicit value $\Psi_e$ calculated using current $\Psi_P$ from Eq. (4.20). The explicit gradient $(\nabla_{\mathbf{n}}\Psi)_e$ is the boundary gradient $\nabla_{\mathbf{n}}\Psi_b$ calculated from $\Psi_N$ in an imaginary neighbour cell by

$$\nabla_{\mathbf{n}}\Psi_b = \frac{C_\Delta}{2}\left[G(\mathbf{R}_N, \Psi_P) - \Psi_P\right], \tag{4.21}$$

where the rotation between cell centres $\mathbf{R}_N = \mathbf{R}_f \bullet \mathbf{R}_f$.

The factor $\mathbf{\Lambda}$ is chosen to minimise the gradient boundary coefficients (see Sec. 4.11). The vector factor is $\mathbf{\Lambda} = \text{diag}\left(\mathbf{I} - \mathbf{R}_N\right)/2$ where "diag" is defined in Sec. 4.12.

For a tensor $\Psi$, the factor is $\mathbf{\Lambda} = \mathbf{\Lambda}_{vec}\mathbf{\Lambda}_{vec}$, where $\mathbf{\Lambda}_{vec}$ are the coefficients for a vector, as described for the symmetry condition in Sec. 4.12.

## Rotation tensor

The rotation tensor $\mathbf{R}$ between two *unit* vectors $\mathbf{n}_1$ and $\mathbf{n}_2$ can be calculated using the Euler-Rodrigues rotation formula,†

$$\mathbf{R} = c\mathbf{I} + (1 - c)\mathbf{k}\mathbf{k} + \mathbf{n}_2\mathbf{n}_1 - \mathbf{n}_1\mathbf{n}_2,$$

where $\mathbf{k} = (\mathbf{n}_1 \times \mathbf{n}_2)/|\mathbf{n}_1 \times \mathbf{n}_2|$ and $c = \mathbf{n}_1 \bullet \mathbf{n}_2$.

†Benjamin Olinde Rodrigues, *De l'attraction des sphéroïdes*, 1815.

## 4.14 Direction mixed condition

In some cases, a boundary condition is sometimes needed which applies a different underlying type — fixed value or gradient — to different components of a non-scalar field.

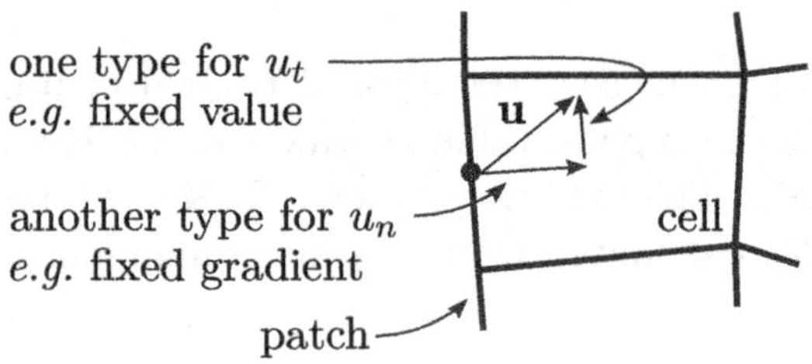


The condition is most readily applied to velocity **u**, where different conditions are applied to its normal and tangential components to the patch, $u_n$ and $u_t$ respectively.

The **direction mixed** condition combines the mixed condition from Sec. 4.9 with the transform conditions of Sec. 4.11. The mixed condition can be first expressed in the form of a general transform condition Eq. (4.13) as

$$\boldsymbol{\Psi}_{\rm f} = v\boldsymbol{\Psi}_{\rm b} + (1 - v)\left[\frac{\nabla_{\mathbf{n}}\boldsymbol{\Psi}_{\rm b}}{C_\Delta} + \boldsymbol{\Psi}_{\rm P}\right]. \tag{4.22}$$

In the direction mixed condition the value fraction $v$ is replaced by a transformation tensor $\mathbf{Y}$, whose components are value fractions in the range 0 to 1, by

$$\boldsymbol{\Psi}_{\rm f} = G(\mathbf{Y}, \boldsymbol{\Psi}_{\rm b}) + G\left([\mathbf{I} - \mathbf{Y}], \left[\frac{\nabla_{\mathbf{n}}\boldsymbol{\Psi}_{\rm b}}{C_\Delta} + \boldsymbol{\Psi}_{\rm P}\right]\right). \tag{4.23}$$

The value fraction tensor $\mathbf{Y}$ is set according to the requirements of the boundary condition which is derived from this direction mixed framework.

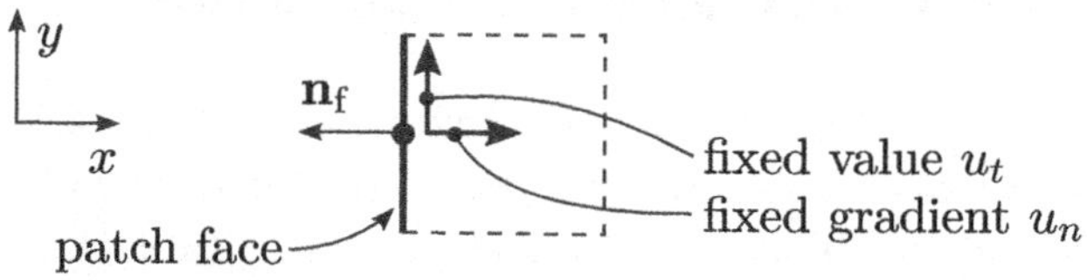


Imagine an example of $\mathbf{u}$ at a face oriented with normal vector $\mathbf{n}_f = (-1, 0, 0)$, for which the $u_n$ condition is fixed gradient and $u_t$ is fixed value. The value fraction must be 1 in the tangential direction and 0 in the normal direction, which gives:

$$\mathbf{Y} = \mathbf{I} - \mathbf{n}_f\mathbf{n}_f = \begin{pmatrix} 0 & 0 & 0 \\ 0 & 1 & 0 \\ 0 & 0 & 1 \end{pmatrix}. \tag{4.24}$$

The boundary condition is then implemented similarly to the symmetry and wedge conditions of Sec. 4.12 and Sec. 4.13. First, $\mathbf{\Psi}_e$ is the calculated $\mathbf{\Psi}_f$ using current $\mathbf{\Psi}_P$ by Eq. (4.23) and the explicit gradient $(\nabla_{\mathbf{n}}\mathbf{\Psi})_e$ is calculated from Eq. (4.17).

The factor $\mathbf{\Lambda}$ corresponds to $\mathbf{Y}$. For a vector field, it is calculated, as in the symmetry condition in Eq. (4.19), by

$$\mathbf{\Lambda} = \operatorname{diag}(\sqrt{\mathbf{Y}}). \tag{4.25}$$

For a tensor $\mathbf{\Psi}$, the factor is $\mathbf{\Lambda} = \mathbf{\Lambda}_{\mathrm{vec}}\mathbf{\Lambda}_{\mathrm{vec}}$, where $\mathbf{\Lambda}_{\mathrm{vec}}$ are the coeffcients for a vector, as described for the symmetry condition in Sec. 4.12.

The condition is implemented using value and gradient factors according to the transform condition, summarised in the table on page 130. Any boundary condition which is based on this direction mixed condition then only requires a description of the $\mathbf{\Psi}_b$, $\nabla_{\mathbf{n}}\mathbf{\Psi}_b$ and $\mathbf{Y}$ parameters.

## 4.15 Inlet-outlet-velocity condition

The inlet-outlet condition was described in Sec. 4.10, which switches between the zero gradient type for outflow and fixed value for inflow. The condition is not so suitable for velocity, *e.g.* at a free boundary, since it is requires an inlet velocity to be *prescribed* when inflow occurs.

The switching based on the flow direction is not a problem in itself. The flow direction comes from the sign of the flux $\phi_{\mathrm{f}}$, as noted in Eq. (4.12), which is determined by the solution of the pressure equation (part of pressure-velocity coupling described in Chapter 5).

The problem is instead that a value of velocity cannot be chosen in the case of inflow since the inflow speed is *determined* by the solution within the domain. This suggests a zero gradient type as a more suitable condition.

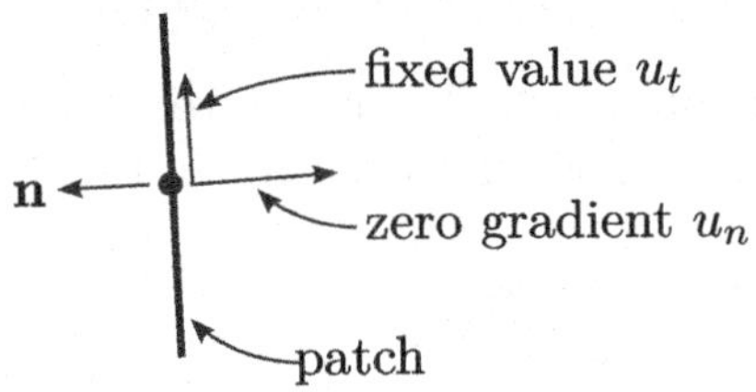


However, Sec. 4.3 discusses that all but one variable should be prescribed at a boundary with inflow. For velocity that variable could be its *normal* component, while a fixed value could still be applied to the *tangential* component.

The direction mixed type described in Sec. 4.14 provides the framework to implement this condition. A reference value $\mathbf{u}_{\mathrm{b}}$ is specified and the reference gradient is zero, *i.e.* $\nabla_{\mathbf{n}}\mathbf{u}_{\mathrm{b}} = 0$.

The value fraction tensor is calculated as

$$\mathbf{Y} = \begin{cases} \mathbf{0} & \text{for outflow,} \\ \mathbf{I} - \mathbf{nn} & \text{for inflow.} \end{cases} \tag{4.26}$$

The fixed value which is applied must be a velocity $\mathbf{u}_t$ tangential to the boundary, which can be calculated by subtracting the normal component from a specified $\mathbf{u}_b$ by $\mathbf{u}_t = \mathbf{u}_b - \mathbf{n}(\mathbf{n} \bullet \mathbf{u}_b)$.

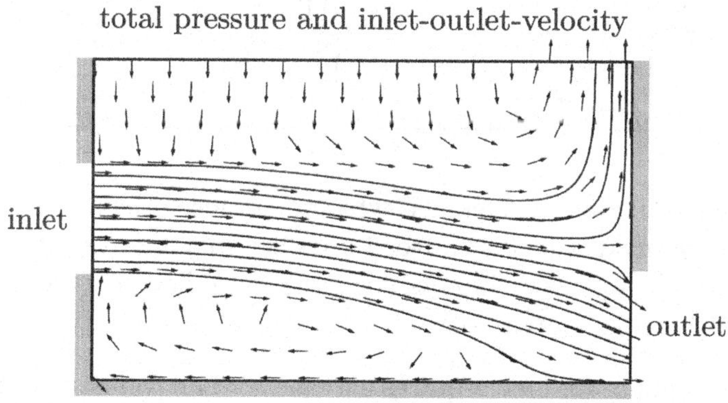


This **inlet-outlet-velocity** condition can be applied at the free boundary in the example from Sec. 4.6. For free boundaries like this, $\mathbf{u}_t = 0$ is the only practical specification, which causes all inflow to be normal to the free boundary, as shown above.

The solution is clearly different to that using $\nabla_{\mathbf{n}}\mathbf{u} = 0$ shown on page 122. There, the inflow direction is determined by the solution rather than prescribed. The solution with $\nabla_{\mathbf{n}}\mathbf{u} = 0$ may be more accurate but the inlet-outlet-velocity condition adheres better to the general principles of boundary conditions, so is likely to be more stable.

It becomes less contentious to set $\mathbf{u}_t = 0$ with the inlet-outlet-velocity condition as the boundary is moved further into the far field, where the flow is more quiescent.

## 4.16 Blended freestream condition

There is a class of problems in CFD that involve external flow around one or more solid bodies, *e.g.* a vehicle, wind turbine, buildings, *etc.*

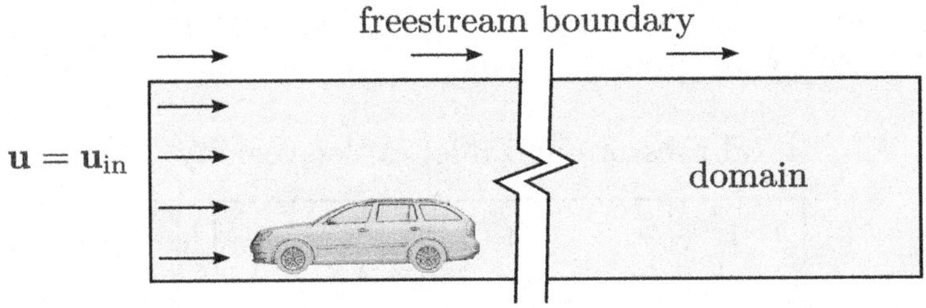


A solution domain is specified which includes the solid body and extends some distance to a free boundary in the *far-field.* A flow velocity $\mathbf{u} = \mathbf{u}_{\text{in}}$is specified which can be applied as a fixed value type at an inlet patch.

The far-field boundary requires attention. The robust conditions at a free boundary for $\mathbf{u}$ and $p$ are inlet-outlet-velocity and total pressure described in Sec. 4.15 and Sec. 4.7 respectively.

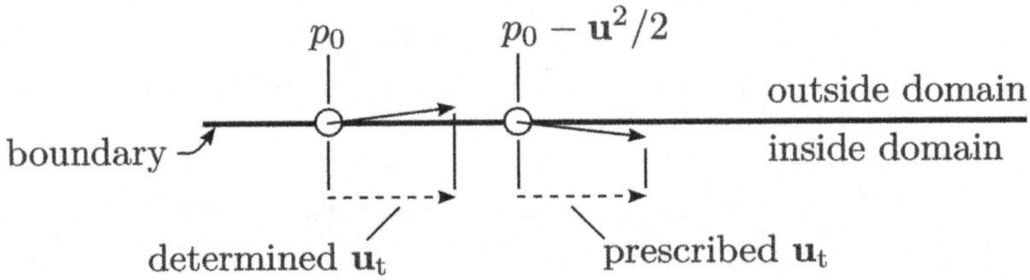


The inlet-outlet-velocity requires $\mathbf{u}_{\text{t}}$ to be prescribed for inflow which may differ significantly from a determined $\mathbf{u}_{\text{t}}$ when there is outflow. Solution accuracy depends on the suitability of the prescribed $\mathbf{u}_{\text{t}}$.

The flow direction in the far-field can often be close to tangential to the boundary, especially with a box-shaped domain. If the flow at one face changes from outflow to inflow, $\mathbf{u}_\mathrm{t}$ suddenly changes to the prescribed fixed value and $p$ decreases by $\mathbf{u}^2/2$. Sometimes a pattern of switching can occur in adjacent faces and repeated switching can slow the convergence of a solution.

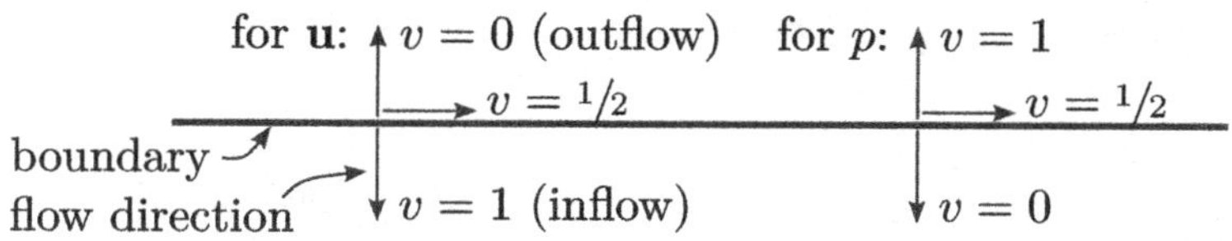


The **blended freestream** condition is a mixed type with zero reference gradient, $\nabla_\mathbf{n}\boldsymbol{\Psi}_\mathrm{b} = 0$, which modifies the value fraction $v$ as shown above. In the limit that the flow direction is normal to the boundary, the condition becomes the fundamental fixed value and zero gradient types for $\mathbf{u}$ and $p$.

Between these extremes, $v$ is blended linearly, *e.g.* for $\mathbf{u}$ by

$$v = \frac{1}{2}\left[1 - \frac{\mathbf{n}\cdot\mathbf{u}}{|\mathbf{u}|}\right], \text{ with } v = \frac{1}{2} \text{ when } |\mathbf{u}| = 0. \tag{4.27}$$

This means $v = 1/2$ for both $\mathbf{u}$ and $p$ for tangential flow.

At a boundary face, $\mathbf{u}$ may be directed normal-inward, causing $v = 1$ by Eq. (4.27). The condition can then "lock" at $v = 1$, so to avoid this, the calculation can use a velocity equating to the mean of the face and neighbour cell value, *i.e.* $\mathbf{u} = (\mathbf{u}_\mathrm{f} + \mathbf{u}_\mathrm{P})/2$.

Note that for $p$, the value fraction is calculated changing the sign of the second term in Eq. (4.27) *i.e.* $v = [1 + (\mathbf{n}\cdot\mathbf{u}/|\mathbf{u})]/2$.

The freestream conditions overcome the problem of switching to improve the convergence of solutions. Boundary velocities are determined, not prescribed, which seems to improve accuracy, *e.g.* in force calculations described in Sec. 8.4 and Sec. 8.6.

## 4.17 External wall heat flux

The boundary condition for temperature (or energy) dictates the heat transfer across a boundary. At a boundary that represents a solid wall, simple conditions can sometimes be applied. However, specialised boundary conditions are often required that control the heat flux across the boundary.

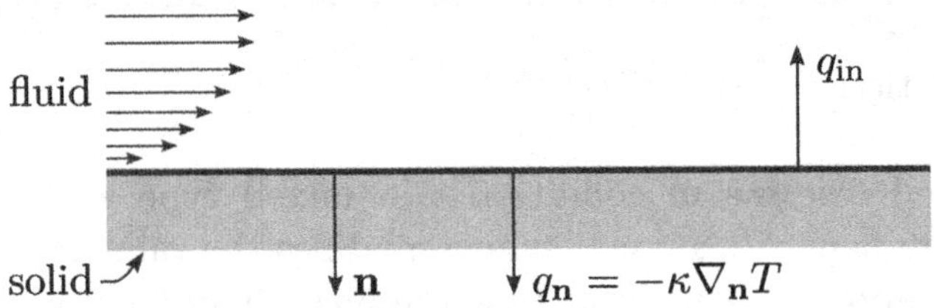


The **fixed temperature** is the simplest condition, setting a fixed value $T = T_{\mathrm{b}}$. This condition provides an approximation for cases for a solid with high *thermal mass*, due to a large mass of material and high conductivity $\kappa$, which helps to maintain constant $T$.

Otherwise the boundary condition sets the heat flux normal to the boundary, $q_{\mathbf{n}}$ derived from Eq. (2.54) by

$$q_{\mathbf{n}} = \mathbf{n} \bullet \mathbf{q} = -\kappa \nabla_{\mathbf{n}} T. \tag{4.28}$$

Another simple condition is zero gradient $\nabla_{\mathbf{n}} T = 0$. This is the **adiabatic** condition, corresponding to zero normal heat flux by Eq. (4.28), suitable when the solid is a thermally insulating material with a large mass and low $\kappa$.

Otherwise a **fixed heat flux** condition specifies an *inward* heat flux $q_{\mathrm{in}}$ as a fixed gradient type with a reference gradient by

$$\nabla_{\mathbf{n}} T_{\mathrm{b}} = q_{\mathrm{in}} / \kappa. \tag{4.29}$$

## Fixed heat transfer coefficient

Another way to specify the heat transfer at an external wall is by Newton's law of cooling.† This general law states the rate of heat loss of a body is directly proportional to the difference between the body temperature $T$ and a surrounding, ambient temperature $T_{\mathrm{a}}$.

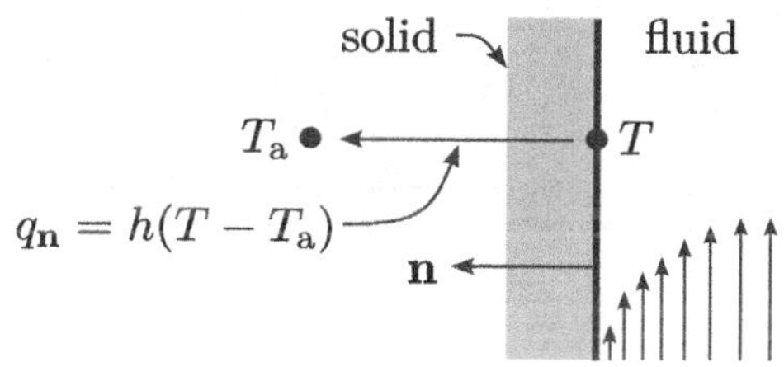


Applied as a boundary condition, $T$ is the fluid temperature at the boundary, and $T_{\mathrm{a}}$ a temperature some distance beyond the solid boundary. A heat transfer coefficient $h$, with SI units $\mathrm{W m^{-2} K^{-1}}$, provides the constant of proportionality such that

$$q_{\mathbf{n}} = h(T - T_{\mathrm{a}}). \tag{4.30}$$

Substituting Eq. (4.28) and rearranging gives an equation for the **fixed heat transfer coefficient** condition:

$$T + (\kappa/h)\nabla_{\mathbf{n}} T = T_{\mathrm{a}}. \tag{4.31}$$

The equation has the form of a Robin condition, Eq. (4.10), so can be implemented as described in Sec. 4.9. The coefficient $h$ is typically characterised for the particular flow regime and solid boundary, by some estimate, experimental measurements or computer simulation.

†Isaac Newton, *Scala graduum caloris. Calorum descriptiones & signa*, Philosophical Transactions, **22**:270, 1701.

## 4.18 Recommended boundary conditions

This chapter covers a range of boundary conditions and their implementations. It first describes a specification of the basic conditions at inlet, outlet and wall boundaries for *subsonic* flow with **fixed value** and **zero gradient**.

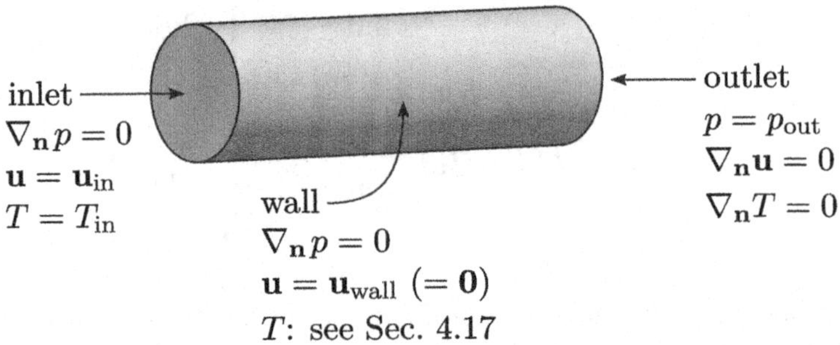


The conditions, based on the propagation of disturbances, are described in Sec. 4.3:

- zero gradient on $p$ at an inlet, fixed value on other variables;
- fixed value on $p$ at an outlet, zero gradient on other variables.

The conditions at a wall are similar to an inlet for $p$ and $\mathbf{u}$, but generally are represented more directly by physical models, *e.g.* the condition for heat flux $q_{\mathbf{n}}$ for $T$.

### Supersonic conditions

The basic conditions for *supersonic* flow are discussed in Sec. 4.5. If the flow speed is supersonic at an inlet, the basic condition is fixed value for $p$; it is zero gradient for $p$ if the flow is supersonic at an outlet.

## Robust, practical conditions

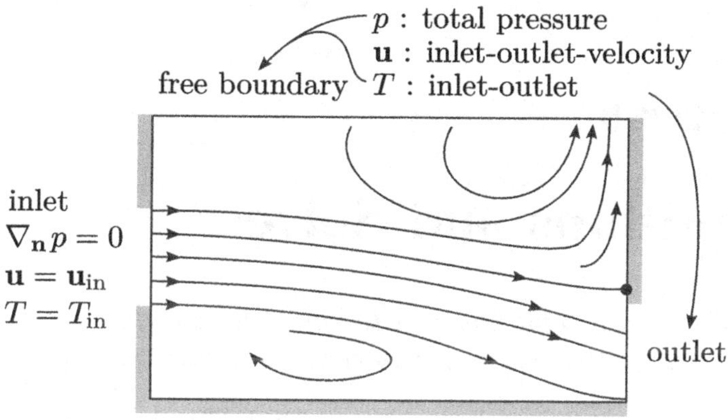


Section 4.6 introduced a free boundary that cannot be defined as an inlet or outlet, but instead often uses the following conditions:

- **total pressure** for $p$, see Sec. 4.7;
- **inlet-outlet-velocity** for $\mathbf{u}$, see Sec. 4.15;
- **inlet-outlet** for $T$, see Sec. 4.10.

These conditions also respond well at an outlet, in the event that some inflow occurs at startup, a rotating structure passes through the boundary *etc.*, see Sec. 4.10.

The **freestream** conditions, Sec. 4.16, are effective for cases with known $\mathbf{u}$ and $p$ at a free, far-field boundary.

The **symmetry** and **wedge** conditions enable suitable cases to be simplified as symmetric and axisymmetric, respectively.

In the presence of a body force $\mathbf{b}$, the zero gradient condition for $p$ at inlets and walls should be replaced by a **fixed gradient** condition $\nabla_{\mathbf{n}} p = \mathbf{n} \bullet \mathbf{b}$, see Sec. 4.4.

# Chapter 5

# Algorithms and Solvers

*"And you run, and you run to catch up with the sun but it's sinking. Racing around to come up behind you again."*

Pink Floyd, *Time* (1973).

The previous chapters describe the components needed to build equations to solve problems in fluid dynamics, which are listed below:

- governing equations and common models, see Chapter 2;
- numerical methods to create discrete sets of linear equations, as matrix equations, see Chapter 3;
- boundary condtitions, which are applied to the matrix equations, see Chapter 4.

The final component is to assemble the set of matrix equations and apply *algorithms* to:

- solve the set of linear equations described by each individual matrix equation;
- combine, or *couple*, solutions from multiple matrix equations to produce the overall solution.

*Fluid equations*, Chapter 2, *e.g.*:

$$\left.\begin{aligned} &\frac{\partial \mathbf{u}}{\partial t} + \nabla \bullet (\mathbf{u}\mathbf{u}) - \nabla \bullet (\nu \nabla \mathbf{u}) = -\nabla p \\ &\nabla^2 p + \nabla \bullet [\nabla \bullet (\mathbf{u}\mathbf{u})] = 0 \end{aligned}\right\}$$

discretisation
Chapter 3

*Boundary conditions*
Chapter 4, *e.g.*:

$p$ : total pressure
$\mathbf{u}$ : inlet-outlet-velocity

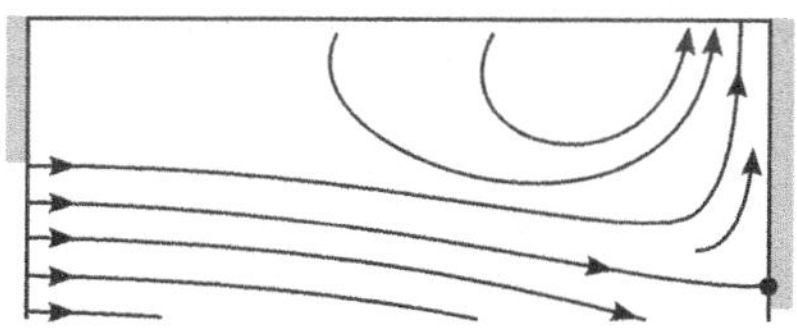

*Matrix equations, e.g.*:

$$\begin{bmatrix} \star & \star & & \star \\ \star & \star & \star & \\ & \star & \star & \star \\ \star & & \star & \star \end{bmatrix} \begin{bmatrix} u_x \end{bmatrix} = \begin{bmatrix} \star \\ \star \\ \star \\ \star \end{bmatrix}$$

$$\begin{bmatrix} \star & \star & & \star \\ \star & \star & \star & \\ & \star & \star & \star \\ \star & & \star & \star \end{bmatrix} \begin{bmatrix} p \end{bmatrix} = \begin{bmatrix} \star \\ \star \\ \star \\ \star \end{bmatrix}$$

*Algorithms and solvers*

This chapter first addresses the solution of individual matrix equations. It presents the most effective algorithms for solving matrix equations encountered with the finite volume method.

The important requirements of these *matrix solvers* are robustness and efficiency, particularly for large simulations with many millions of cells. Solvers than accompany the finite volume method are generally *iterative* in nature, in which the solution converges towards a prescribed acceptable level of accuracy, or *tolerance*, over a number of iterations.

The principal algorithms that couple sets of matrix equations will then be presented. The coupling of mass and momentum conservation with equations for $\mathbf{u}$ and $p$ is challenging, and so requires particular attention.

## 5.1 Structure of matrices

As described in Sec. 3.4, a matrix equation for solution variable $\boldsymbol{\Psi}$ contains of a set of coefficients $a_{i,j}$ where each row $i$ corresponds to the linear equation for the cell with index $i$ as follows:

$$\begin{bmatrix} a_{1,1} & a_{1,2} & a_{1,3} & \cdots & a_{1,N} \\ a_{2,1} & a_{2,2} & a_{2,3} & \cdots & a_{2,N} \\ a_{3,1} & a_{3,2} & a_{3,3} & \cdots & a_{3,N} \\ \vdots & \vdots & \vdots & \ddots & \vdots \\ a_{N,1} & a_{N,2} & a_{N,3} & \cdots & a_{N,N} \end{bmatrix} \begin{bmatrix} \boldsymbol{\Psi}_1 \\ \boldsymbol{\Psi}_2 \\ \boldsymbol{\Psi}_3 \\ \vdots \\ \boldsymbol{\Psi}_N \end{bmatrix} = \begin{bmatrix} b_1 \\ b_2 \\ b_3 \\ \vdots \\ b_N \end{bmatrix} \tag{5.1}$$

### Matrix sparsity

Imagine creating a matrix equation for a transport equation for a scalar field, *e.g.* Eq. (2.65) with zero heat source $r = 0$.

$$\frac{\partial T}{\partial t} + \nabla \bullet (\mathbf{u}T) - \nabla \bullet (\alpha \nabla T) = 0$$

discretise + boundary conditions

$\rightarrow [\mathbf{A}][T] = [\mathbf{b}]$

$\rightarrow$ matrix $[\mathbf{A}]$ is sparse

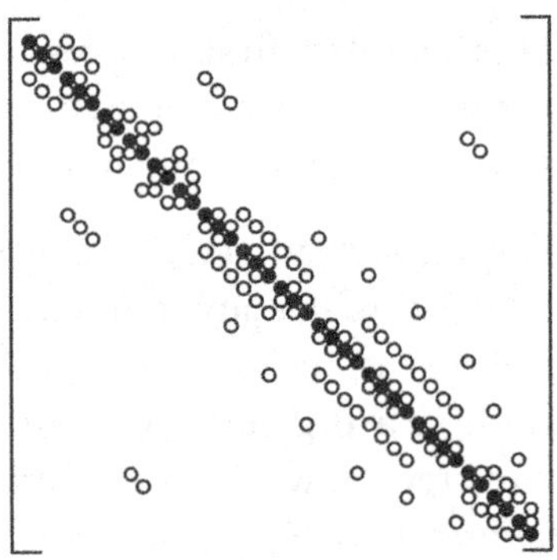

The figure shows a matrix of size $N = 40$, where $N$ is the number of cells. Circles denote non-zero coefficients, which are filled for the diagonal coefficients ($a_{i,i}$). The matrix $[\mathbf{A}]$ is *sparse*, *i.e.* the majority of the coefficients $a_{i,j}$ are 0 (zero).

The sparsity is due to each cell interacting only with adjacent cells connected through its faces. For example, a 3D mesh of hexahedral cells produces up to 7 coefficients per matrix row, with one diagonal coefficient corresponding to a particular cell and 6 off-diagonal coefficients for the neighbour cells.

## Matrix (a)symmetry

A symmetric matrix possesses the same coefficients across the diagonal, *i.e.* $a_{i,j} = a_{j,i}$. The discretisation of a Laplacian term, *e.g.* $\nabla \bullet \alpha \nabla T$, described in Sec. 3.7, produces coefficients that are symmetric because the surface normal gradient Eq. (3.5) uses the current and neighbour cell values in equal measure.

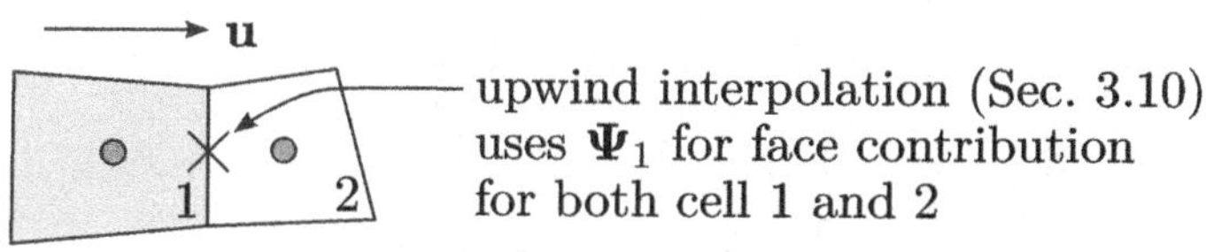


However, the discretisation of an *advection* term, *e.g.* $\nabla \bullet (\mathbf{u}T)$, generally produces *asymmetric* coefficients. For example, if the upwind scheme is applied for flow from cell 1 into cell 2, then there would a contribution to $a_{2,1}$ but not to $a_{1,2}$.

## Matrix size

Parallel simulation allows affordable solution on huge meshes. For a mesh of size $N = 10^6$, the matrix would have $N^2 = 10^{12}$ coefficients, typically with $7 \times 10^6$ that are are non-zero.

For reasons of efficiency, zero coefficients are not stored in the computer's memory. Instead the storage provides an array of non-zero coefficients and addressing arrays of the corresponding row and column indices for each coefficient.

## 5.2 Gauss-Seidel method

Finite volume numerics generally uses *iterative* methods to solve each matrix equation. These methods calculate approximate solutions for $\boldsymbol{\Psi}$, which become more accurate with successive iterative solutions.

Iterative methods are preferred because they are more efficient than direct methods, which solve a matrix equation exactly. Gaussian elimination, which is the numerical basis for direct solution methods, has a computational cost $\propto N^3$. This is prohibitive for many sizes of mesh in finite volume CFD.

**Gauss-Seidel**[†] is a simple, iterative method which is generally effective for solving transport equations such as the example in Sec. 5.1. The method is illustrated by a sample equation

$$\begin{bmatrix} 3 & -1 & -1 \\ -1 & 5 & -2 \\ -1 & -2 & 4 \end{bmatrix} \begin{bmatrix} \boldsymbol{\Psi}_1 \\ \boldsymbol{\Psi}_2 \\ \boldsymbol{\Psi}_3 \end{bmatrix} = \begin{bmatrix} 6 \\ -2 \\ 3 \end{bmatrix}. \tag{5.2}$$

The equations can be rewritten by: a) multiplying the off-diagonal coefficients by $\boldsymbol{\Psi}$; b) subtracting the result from the right hand side (r.h.s.); c) and, dividing by the diagonal coefficient. *i.e.*:

$$\boldsymbol{\Psi}_1 := \frac{1}{3}\left(6 + \boldsymbol{\Psi}_2 + \boldsymbol{\Psi}_3\right); \tag{5.3a}$$

$$\boldsymbol{\Psi}_2 := \frac{1}{5}\left(-2 + \boldsymbol{\Psi}_1 + 2\boldsymbol{\Psi}_3\right); \tag{5.3b}$$

$$\boldsymbol{\Psi}_3 := \frac{1}{4}\left(3 + \boldsymbol{\Psi}_1 + 2\boldsymbol{\Psi}_2\right). \tag{5.3c}$$

Starting with, $\boldsymbol{\Psi}_1 = \boldsymbol{\Psi}_2 = \boldsymbol{\Psi}_3 = 0$, new values of $\boldsymbol{\Psi}$ are calculated by Eq. (5.3a), Eq. (5.3b) and Eq. (5.3c) in sequence, where the notation "$a := b$" denotes "$a$ is assigned the value of $b$".

---

[†] Carl Friedrich Gauss, *letter to Christian Gerling*, 1823.

The first solution of Eq. (5.3a) is $\boldsymbol{\Psi}_1 = 2$. The updated $\boldsymbol{\Psi}_1$ is substituted in Eq. (5.3b), whose solution is $\boldsymbol{\Psi}_2 = 0$. Both updated values are substituted in Eq. (5.3c) to give $\boldsymbol{\Psi}_3 = 1.25$.

The process is then repeated and through successive sweeps over the equations the solution converges as shown below.

| Variable | Start | Sweep 1 | ...2 | ...3 | ...4 |
|---|---|---|---|---|---|
| $\boldsymbol{\Psi}_1$ | 0.0000 | 2.0000 | 2.4167 | 2.7431 | 2.8821 |
| $\boldsymbol{\Psi}_2$ | 0.0000 | 0.0000 | 0.5833 | 0.8069 | 0.9121 |
| $\boldsymbol{\Psi}_3$ | 0.0000 | 1.2500 | 1.6458 | 1.8392 | 1.9266 |

| Variable | Sweep 5 | ...6 | ...7 | ...8 | ...9 |
|---|---|---|---|---|---|
| $\boldsymbol{\Psi}_1$ | 2.9462 | 2.9755 | 2.9888 | 2.9949 | 2.9977 |
| $\boldsymbol{\Psi}_2$ | 0.9599 | 0.9817 | 0.9916 | 0.9962 | 0.9983 |
| $\boldsymbol{\Psi}_3$ | 1.9665 | 1.9847 | 1.9930 | 1.9968 | 1.9985 |

The error is $\epsilon_i = \boldsymbol{\Psi}_i - (\boldsymbol{\Psi}_i)_{\text{ex}}$, *i.e.* the difference between the approximate and exact values $(\boldsymbol{\Psi}_1)_{\text{ex}} = 3$, $(\boldsymbol{\Psi}_2)_{\text{ex}} = 1$ and $(\boldsymbol{\Psi}_3)_{\text{ex}} = 2$. After 9 sweeps $\epsilon_i \approx 0.002$ for all variables, *i.e.* within 0.2% of the exact solution.

In summary, the Gauss-Seidel method is the following sequence of calculations for $i = 1, 2, \ldots, N$, repeated until convergence:

$$\boldsymbol{\Psi}_i := \frac{1}{a_{i,i}} \left( b_i - \sum_{\substack{j=1 \\ i \neq j}}^{N} a_{i,j} \boldsymbol{\Psi}_j \right). \tag{5.4}$$

The Gauss-Seidel method, *applied to sparse matrices*, has a computational cost $\propto N$, making it practical for simulations with large meshes that often occur in finite volume CFD.

Convergence of the method, and convergence measures for iterative methods in general, are discussed in the following sections.

## 5.3 Convergence

The Gauss-Seidel method was demonstrated in Sec. 5.2 using a sample problem, Eq. (5.2), that converged to within 0.2% accuracy in 9 solution sweeps. This section discusses the criteria for convergence of a solution.

The easiest explanation of convergence considers the error $\epsilon_i$ for each equation $i$, defined in Sec. 5.2. Substituting $\mathbf{\Psi}_i = (\mathbf{\Psi}_i)_{\text{ex}}+\epsilon_i$ in each term in Eq. (5.3), *e.g.* in Eq. (5.3a), gives

$$(\mathbf{\Psi}_1)_{\text{ex}} + \epsilon_1 = \frac{1}{3}\left(6 + (\mathbf{\Psi}_2)_{\text{ex}} + \epsilon_2 + (\mathbf{\Psi}_2)_{\text{ex}} + \epsilon_3\right). \tag{5.5}$$

Since $3(\mathbf{\Psi}_1)_{\text{ex}} - (\mathbf{\Psi}_2)_{\text{ex}} - (\mathbf{\Psi}_3)_{\text{ex}} \equiv 6$, Eq. (5.5) reduces to

$$\epsilon_1 = \frac{1}{3}\epsilon_2 + \frac{1}{3}\epsilon_3. \tag{5.6}$$

The magnitude of error $|\epsilon_1|$ is *at least* as large as the sum of the terms on the r.h.s., but is smaller if the signs of $\epsilon_2$ and $\epsilon_3$ are different, *i.e.*

$$|\epsilon_1| \leq \frac{1}{3}|\epsilon_2| + \frac{1}{3}|\epsilon_3|. \tag{5.7}$$

Repeating for Eq. (5.3b) and Eq. (5.3c) gives:

$$|\epsilon_2| \leq \frac{1}{5}|\epsilon_1| + \frac{2}{5}|\epsilon_3|; \tag{5.8a}$$

$$|\epsilon_3| \leq \frac{1}{4}|\epsilon_1| + \frac{1}{2}|\epsilon_2|. \tag{5.8b}$$

The solution begins with an initial error of $\epsilon_1 = 6$, $\epsilon_2 = -2$ and $\epsilon_3 = 3$. After one sweep the error is $\epsilon_1 = 0.33$, $\epsilon_2 = 1.53$ and $\epsilon_3 = 0.85$.

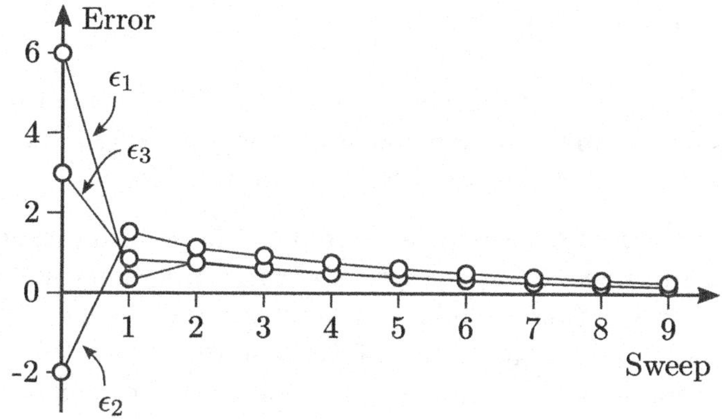


The error is quickly distributed evenly, such that $\epsilon_1$ and $\epsilon_3$ are almost identical at sweep 2. The errors continue to reduce since Eq. (5.7) and Eq. (5.8) guarantee that no error is greater than the average of the other errors.

## Condition for convergence

The behaviour of this problem indicates a convergence condition for the Gauss-Seidel method: the magnitude of the diagonal coefficient in each matrix row must be greater than or equal to the sum of the magnitudes of the other coefficients in the row; in one row at least, the "greater than" condition must hold.

This is known as *diagonal dominance*, which is a *sufficient* condition for convergence, described mathematically as

$$|a_{i,i}| \geq \sum_{\substack{j=1 \\ i \neq j}}^{N} |a_{i,j}| \quad \text{for all } i, \tag{5.9}$$

where the '>' condition must be satisfied for at least one $i$. The description of the condition as "sufficient" means that convergence may occur when the condition is not met.

## 5.4 Residual

In Sec. 5.3, we established a criterion for convergence of the Gauss-Seidel method. We now need a way to estimate a *level* of convergence to determine when to stop iterating.

The analysis of convergence centred on the solution error $\epsilon_i$, introduced in Sec. 5.2. In practice, $\epsilon_i$ cannot be determined since the exact solution is unknown. Instead the *residual* provides a measure of the accuracy of the solution. The *residual vector* **r** represents the change to the solution of the equation, required to make $\boldsymbol{\Psi}$ exact, according to

$$\begin{aligned}\mathbf{r} &= \mathbf{A} \bullet \boldsymbol{\Psi}_{\text{ex}} - \mathbf{A} \bullet \boldsymbol{\Psi} \\ &= \mathbf{b} - \mathbf{A} \bullet \boldsymbol{\Psi},\end{aligned} \tag{5.10}$$

where $\boldsymbol{\Psi}_{\text{ex}} = (\boldsymbol{\Psi}_i)_{\text{ex}}$.

> For the **remainder of this chapter**, the matrix notation $[\mathbf{A}][\boldsymbol{\Psi}] = [\mathbf{b}]$ is replaced by $\mathbf{A} \bullet \boldsymbol{\Psi} = \mathbf{b}$, equivalent to vector notation with an $N \times N$ tensor **A**.

The vector **r** (of size $N$) provides *one value per matrix row*, with both positive and negative values. A measure of residual given by a *single* value, is defined as

$$r = \frac{\|\mathbf{b} - \mathbf{A} \bullet \boldsymbol{\Psi}\|}{\|\mathbf{A} \bullet \boldsymbol{\Psi} - \mathbf{A} \bullet \overline{\boldsymbol{\Psi}}\| + \|\mathbf{b} - \mathbf{A} \bullet \overline{\boldsymbol{\Psi}}\|}, \tag{5.11}$$

where $\|\ldots\|$ is the matrix norm, calculated as the sum of the magnitude of each component, *e.g.* $\|\mathbf{b}\| = \sum_{i=1}^{N} |b_i|$; the mean value of $\boldsymbol{\Psi}$ over all cells is denoted by $\overline{\boldsymbol{\Psi}}$.

The residual $r$ provides a measure of error in the *solution* of $\mathbf{A} \bullet \boldsymbol{\Psi} = \mathbf{b}$, rather than the *absolute error* $\epsilon_i$. It is divided by the norms of $\mathbf{A} \bullet \boldsymbol{\Psi}$ and **b** to reduce its dependency on the scale of the geometry and solution variable. By reducing its scale-dependency, $r$ can be used to compare the level of error equitably between simulations at different scales.

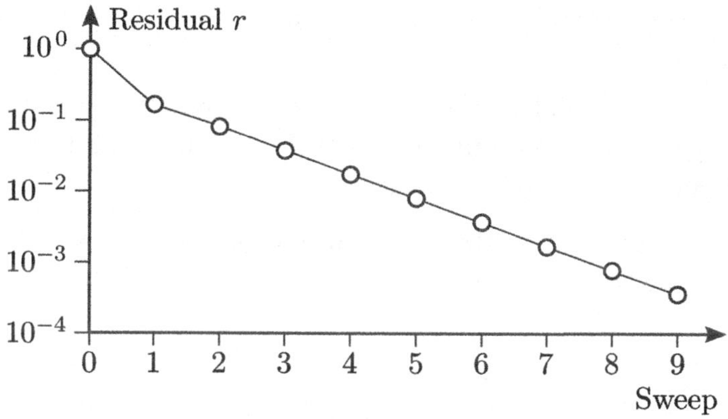


The figure above shows $r$ calculated from Eq. (5.11) following successive sweeps of the Gauss-Seidel method (starting from the initial $r = 1$). The graph uses a logarithmic vertical scale since the values of $r$ extend over 4 orders of magnitude.

## Tolerance

CFD software generally provides the following controls to stop the iterative solver:

- *absolute* tolerance $r_{\text{abs}}$ ;
- *relative* tolerance $r_{\text{rel}}$ .

Sweeping ceases if *either* tolerance condition is satisfied: $r < r_{\text{abs}}$; or $r < r_{\text{rel}} \times r_0$, where $r_0$ is the initial residual within the particular solution step. The $r_{\text{rel}}$ criterion is often deactivated by setting $r_{\text{rel}} = 0$, especially for transient simulations when sufficient accuracy is required at every solution step.

## 5.5 Diagonal dominance

Let us return to the transport equation in Sec. 5.1

$$\frac{\partial T}{\partial t} + \nabla \cdot (\mathbf{u}T) - \nabla \cdot (\alpha \nabla T) = 0 \tag{5.12}$$

If the Gauss-Seidel method is applied to solve the equation, a convergent solution is only *guaranteed* if the diagonal dominance condition of Sec. 5.3 is satisfied.

The contributions to the diagonal and off-diagonal coefficients are summarised below for each term in the equation.

| $\dfrac{\partial T}{\partial t}$ | $+\nabla \cdot (\mathbf{u}T)$ | $-\nabla \cdot (\alpha \nabla T)$ |
|---|---|---|
| $\dfrac{V}{\Delta t}\boldsymbol{\Psi}_{\mathrm{P}}$ | $\sum_{\mathrm{f}} \phi_{\mathrm{f}} (w\, \boldsymbol{\Psi}_{\mathrm{P}}$ $\ldots + (1-w)\, \boldsymbol{\Psi}_{\mathrm{N}})$ | $\sum_{\mathrm{f}} \lvert \mathbf{S}_{\mathrm{f}} \rvert\, \alpha_{\mathrm{f}}\, C_{\Delta} (\boldsymbol{\Psi}_{\mathrm{P}} - \boldsymbol{\Psi}_{\mathrm{N}})$ |
| **diagonal only** | **generally not diagonally equal** | **diagonally equal** |
| **symmetric** | **asymmetric** | **symmetric** |

The Laplacian term $-\nabla \cdot (\alpha \nabla T)$ (note the sign), is discretised by Eq. (3.2) and Eq. (3.5), providing a positive contribution to the diagonal coefficient (relating to $\boldsymbol{\Psi}_{\mathrm{P}}$) equal to the sum of the magnitude of negative off-diagonal coefficients (for $\boldsymbol{\Psi}_{\mathrm{N}}$).

The contribution from the advection term $\nabla \cdot (\mathbf{u}T)$, described by Eq. (3.8), is demonstrated above using the linear interpolation scheme of Eq. (3.4). The sign of $\phi_{\mathrm{f}}$ depends on whether the flux in incoming or outgoing from the cell. Some contributions to off-diagonal coefficients are thus negative and some positive, with the diagonal contribution tending to zero as the mesh becomes more regular and flow becomes more uniform.

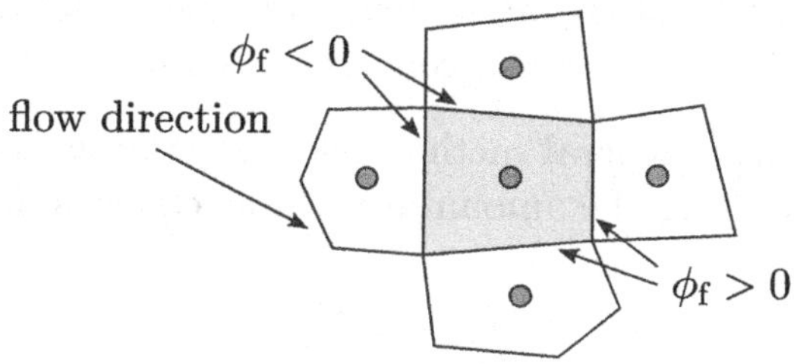


Advection with linear interpolation maintains diagonal equality while its positive off-diagonal contributions offset the negative contributions from the Laplacian term.

But once the advection contributions exceed the magnitude of the Laplacian contributions, diagonal equality is **not** achieved. This occurs when $\text{Pe} > 1/(1-w)$ ($= 2$ for a regular mesh), where the "numerical" Péclet number $\text{Pe} = \phi_f/(|\mathbf{S}_f|\, \alpha_f\, C_\Delta)$ is specified at cell faces from its definition in Sec. 2.21 with $L = C_\Delta^{-1}$.

Diagonal equality is achieved with *upwind interpolation* and any scheme that provides an explicit correction on upwind, *e.g.* linear upwind. This is conditional on mass conservation being satisfied, as discussed in Sec. 3.22. If mass is not conserved, the bounded advection scheme, described in that section, is required to ensure diagonal equality.

The time derivative $\partial T/\partial t$ increases the diagonal coefficient only, so promotes diagonal dominance. Notably, the diagonal contribution of $V/\Delta t$ from $\partial T/\partial t$ becomes larger than $\sum_f \phi^+$ from advection with upwind when $\text{Co} < 1$.

In summary, *diagonal dominance is not guaranteed* due to the contribution of the advection term.

## 5.6 Under-relaxation

Sec. 5.5 concludes that the matrix of a typical transport equation is not guaranteed to be diagonally dominant. Some action may therefore be required to ensure a convergent solution.

Under-relaxation is a general method used to improve solution convergence by limiting the amount a variable changes during a solution step.

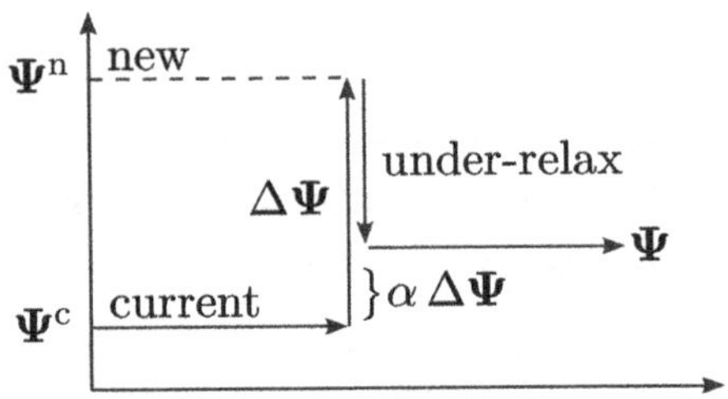


During a solution step, assume a single value of a field $\Psi$ in one cell changes from its current value $\Psi^c$ to the new value $\Psi^n$. Under-relaxation would limit the change $\Delta\Psi = \Psi^n - \Psi^c$ by a fraction $\alpha$, $0 < \alpha \leq 1$, so that the value taken from that solution step is

$$\begin{aligned}\Psi &= \Psi^c + \alpha\,\Delta\Psi \\ &= (1-\alpha)\Psi^c + \alpha\Psi^n.\end{aligned} \tag{5.13}$$

In some situations, Eq. (5.13) is applied after a solution step. This simple approach is known as *field under-relaxation*, which has one notable disadvantage that it requires additional storage of the intermediate field $\Psi^n$ in computer memory.

When a solution step involves solving a matrix equation, the new values $\Psi^n$ come from an iterative method like Gauss-Seidel. Combining the under-relaxation of Eq. (5.13) with the Gauss-

Seidel calculation of Eq. (5.4) gives:

$$\mathbf{\Psi}_i = (1-\alpha)\mathbf{\Psi}_i^{\mathrm{c}} + \frac{\alpha}{a_{i,i}}\left( b_i - \sum_{\substack{j=1 \\ i \neq j}}^{N} a_{i,j}\mathbf{\Psi}_j \right). \tag{5.14}$$

Rearranging Eq. (5.14) gives the following relation:

$$\frac{1}{\alpha}\, a_{i,i}\mathbf{\Psi}_i + \sum_{\substack{j=1 \\ i \neq j}}^{N} a_{i,j}\mathbf{\Psi}_j = b_i + \left(\frac{1}{\alpha} - 1\right) a_{i,i}\mathbf{\Psi}_i^{\mathrm{c}}. \tag{5.15}$$

Equation 5.15 is simply the matrix equation $\mathbf{A} \bullet \mathbf{\Psi} = \mathbf{b}$ modified by:

- increasing the diagonal coefficients $a_{i,i}$ by division by $\alpha$;
- multiplying the difference between the new and original $a_{i,i}$ coefficients by the current $\mathbf{\Psi}^{\mathrm{c}}$ and adding it to the source $\mathbf{b}$.

Modifying the matrix equation this way, known as *equation under-relaxation*, provides an alternative to Eq. (5.13) for under-relaxing a solution of $\mathbf{\Psi}$, without the temporary storage of $\mathbf{\Psi}^{\mathrm{n}}$.

## Ensuring diagonal dominance

The modification to $\mathbf{A} \bullet \mathbf{\Psi} = \mathbf{b}$ expressed by Eq. (5.15) inspires a strategy to ensure diagonal dominance of the matrix as follows.

Each diagonal coefficient which does not satisfy Eq. (5.9) is increased until it is diagonally equal. The change to the coefficient is multiplied by the current $\mathbf{\Psi}^{\mathrm{c}}$ and added to $\mathbf{b}$.

This approach to ensure diagonal dominance is effective since it only modifies matrix coefficients where necessary. Otherwise, if the discretisation schemes, and Pe and Co are favourable, then no changes to the matrix are necessary.

## 5.7 Iterative solution

Imagine a problem with steady flow in which $T$ is changing at a boundary, *e.g.* at an inlet. The system could be simulated by solving Eq. (5.12) with a $\mathbf{u}$ field which is constant in time.

The simulation would evolve by an iterative sequence which increases time $t$ by an increment $\Delta t$, then solves the equation, as shown below.

**START** $t = 0$

↓

**WHILE** $t < t_{\text{end}}$ ← Loop

↓

$t := t + \Delta t$

↓

Solve for $T$: $\dfrac{\partial T}{\partial t} + \nabla \boldsymbol{\cdot} (\mathbf{u}T) - \nabla \boldsymbol{\cdot} (\alpha \nabla T) = 0$

### Setting tolerances

The $T$ equation is solved using an iterative method such as Gauss-Seidel. Tolerance parameters determine when the method stops iterating within the current solution step.

To capture the transient solution for $T$ accurately, the equation would need to converge to a suitable absolute tolerance $r_{\text{abs}}$ at every time step. The question is: what $r_{\text{abs}}$ is suitable to give an accurate solution? — without setting it to a very small value, *e.g.* $10^{-15}$, which would would be highly inefficient, due to an excessive number of solver iterations.

Solution accuracy should be determined by some measure, or *metric*, *e.g.* temperature rise, pressure drop, drag coefficient, *etc.*, relevant to the *objective* of the simulation.

The metric should be initially calculated using a value of $r_{\text{abs}}$, *e.g.* $10^{-3}$. It can then be calculated again at a lower $r_{\text{abs}}$, typically decreasing by one order of magnitude, *i.e.* $10^{-4}$.

The metrics are then compared. If the difference is within the required level of accuracy, then $r_{\text{abs}}$ is optimal. Otherwise, $r_{\text{abs}}$ would need to be decreased further and the process repeated.

## Residual calculation at intervals

The residual calculation of Eq. (5.11) incurs a significant computational cost since it involves multiplications of matrices. That cost can be reduced by setting an interval on the *number of sweeps* between residual calculations.

The downside is that convergence checks can only happen when the residual is calculated. If the residual is calculated at intervals of every second sweep, a solution that would otherwise converge at an odd number of sweeps is prolonged by a extra sweep.

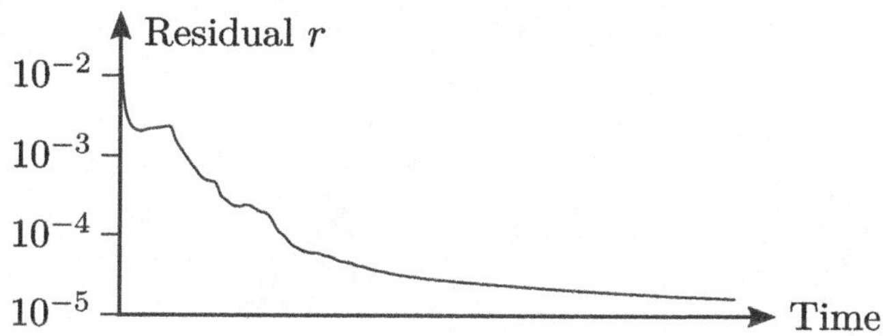


Residuals are high when a simulation starts so a relatively high number of sweeps, *e.g.* 10, may be needed for convergence. The saving of calculating residuals at an interval may then offset the additional cost of unnecessary additional sweeps, giving a lower net cost per time step. However, as the residuals decrease and fewer sweeps are needed, attempts to reduce cost by intermittent residual calculation can be ineffective or counter-productive.

## 5.8 Accelerating convergence

The Gauss-Seidel method sweeps across cells, calculating the changes in values of fields, *e.g.* $T$. Each sweep propagates the changes in $T$ through cells in the domain in much the same way that the flow transports $T$ physically.

The method is therefore remarkably effective, so the scope to accelerate convergence to reduce computational cost is limited. Two options that may reduce cost, relating to mesh numbering, are described in this section.

### Symmetric Gauss-Seidel

The Gauss-Seidel method sweeps through equations in sequence $i = 1, 2, \ldots, N$ of cell numbering. The changes in the solution "flow" in a direction, based on the choice of cell numbering which may be arbitrary and therefore not optimal.

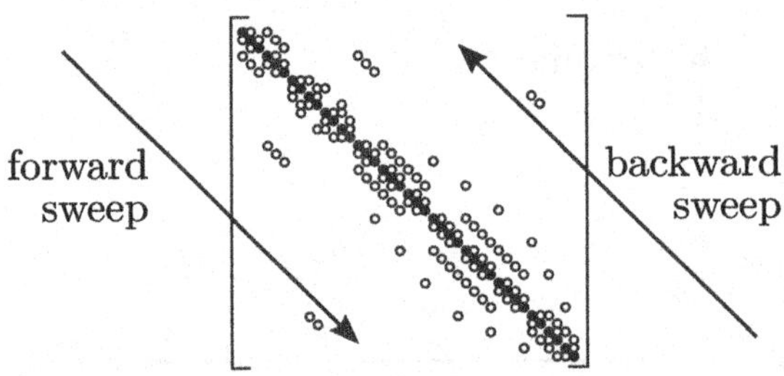


The **symmetric Gauss-Seidel** method follows the standard $i = 1, 2, \ldots, N$ sequence initially, known as a *forward* sweep. For every second sweep, the sequence is executed in reverse, *i.e.* $i = N, N-1, \ldots, 1$, known as a *backward* sweep.

The change in sweep direction moderates the directional bias of the standard Gauss-Seidel method which may reduce the total iterations for convergence and reduce cost.

## Mesh numbering

When a matrix equation is solved in sequence $i = 1, 2, \ldots, N$, values from one row of the equation are substituted in subsequent equations. Convergence improves if the substitutions are made quickly, *e.g.* in the subsequent 2 or 3 equations rather than in the last equation.

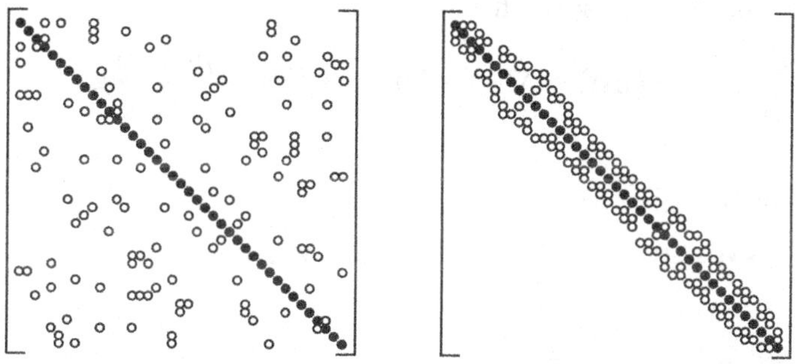

random numbering

Reverse Cuthill-McKee renumbering

Convergence is therefore affected by the cell numbering which determines the position of non-zero coefficients in the matrix. If the numbering is approaching random, *e.g.* as a consequence of the mesh generation, convergence is sub-optimal.

Where this occurs, a method such as reverse Cuthill-McKee[†] can be applied to renumber the cells of a mesh to cluster non-zero coefficients around the diagonal, as shows above.

The matrix in Sec. 5.1 is from a mesh generated in a structured manner with ordered indexing along rows of cells. Renumbering a matrix like that generally provides little benefit since it already has many non-zero coefficients clustered around the diagonal.

[†] Elizabeth Cuthill and James McKee, *Reducing the bandwidth of sparse symmetric matrices*, 1969; "reversed" by Alan George and Joseph Liu, *Computer Solution of Large Sparse Positive Definite Systems*, 1981.

## 5.9 Systems of equations

Most CFD calculations involve solving a system of equations that represent the physics of the problem. For example, laminar flow by natural convection can be represented by the equations introduced in Sec. 2.20, reproduced below.

**momentum**: solve for $\mathbf{u}$

$$\frac{\partial \mathbf{u}}{\partial t} + \nabla \bullet (\mathbf{u}\mathbf{u}) - \nabla \bullet (\nu \nabla \mathbf{u}) = -\nabla p + \frac{\rho(T, p)}{\rho_0}\mathbf{g}$$

**mass**: solve for $p$

$$\nabla \bullet \mathbf{u} = 0 \longrightarrow \nabla^2 p + \nabla \bullet [\nabla \bullet (\mathbf{u}\mathbf{u})] = 0$$

**energy**: solve for $T$

$$\frac{\partial T}{\partial t} + \nabla \bullet (\mathbf{u}T) - \nabla \bullet (\alpha \nabla T) = r$$

The system provides 3 equations (1 vector, 2 scalar) which can be solved for 3 unknowns, $\mathbf{u}$, $p$ and $T$.

As discussed in Sec. 3.4, the finite volume method solves an individual matrix equation for each variable, *e.g.* $\mathbf{A} \bullet T = \mathbf{b}$ for $T$. The vector equation for $\mathbf{u}$ is decoupled into 3 matrix equations for individual components, *i.e.* $\mathbf{A} \bullet u_x = \mathbf{b}$, $\mathbf{A} \bullet u_y = \mathbf{b}$ and $\mathbf{A} \bullet u_z = \mathbf{b}$.

Each individual matrix equation for one solution variable, *e.g.* $u_x$, incorporates terms from other variables, *e.g.* $\nabla p$, into the source vector $\mathbf{b}$. The contribution to $\mathbf{b}$ is calculated using current values of the respective variables. Systems of equations are thereby solved by *successive substitution* of solved variables into the source vectors of subsequent equations.

An iterative solution for a single equation, like the one in Sec. 5.7, can be extended to a system of equations. Time $t$ is incremented by $\Delta t$ and equations are solved in sequence, before returning to start the next time step with the increment of $t$.

The substitutions in the momentum and pressure equations are particularly important, culminating in *corrections* to $\mathbf{u}$ and the advective flux $\phi_\mathrm{f}$, discussed in Sec. 5.10.

$$\frac{\partial \mathbf{u}}{\partial t} + \nabla \cdot (\underbrace{\mathbf{u}\mathbf{u}}_{\text{uses } \phi_\mathrm{f}}) - \nabla \cdot (\nu \nabla \mathbf{u}) = -\nabla p + \frac{\rho(T,p)}{\rho_0}\mathbf{g}$$

$\mathbf{A} \cdot \mathbf{u} = \mathbf{b}$ : solve for $\mathbf{u}$

↓ substitute $\mathbf{u}$

$$\nabla^2 p + \nabla \cdot [\nabla \cdot (\underbrace{\mathbf{u}\mathbf{u}}_{\text{uses } \phi_\mathrm{f}})] = 0$$

$\mathbf{A} \cdot p = \mathbf{b}$ : solve for $p$

↓ substitute $p$ to *correct* $\phi_\mathrm{f}$ and $\mathbf{u}$

At the start of any time step the current $\mathbf{u}$ becomes $\mathbf{u}^\mathrm{o}$ for the discretisation of the $\partial \mathbf{u}/\partial t$ term in the momentum equation. The advection term $\nabla \cdot (\mathbf{u}\mathbf{u})$ is discretised by Eq. (3.8), treating one $\mathbf{u}$ as flux $\phi_\mathrm{f}$ and the other as the advected quantity.

The equation is solved for $\mathbf{u}$. The new solution for $\mathbf{u}$ is substituted into the $p$ equation which is solved for $p$. The new solution for $p$ is then used to *correct* $\phi_\mathrm{f}$ in order to help enforce the mass conservation constraint $\sum_\mathrm{f} \phi_\mathrm{f} = 0$ ($\nabla \cdot \mathbf{u} = 0$).

Before the current solution step is completed, $\mathbf{u}$ is also corrected to reduce the error in the discretisation of $\partial \mathbf{u}/\partial t$ when it then becomes $\mathbf{u}^\mathrm{o}$ in the following solution step. The correction also provides a better "initial guess" $\mathbf{u}$ for the matrix solution of the next momentum equation, which reduces the solution time.

## 5.10 Pressure-velocity coupling

The previous section combined equations for $\mathbf{u}$ and $p$, governing momentum and mass conservation, in a sequential solution.

The algorithms used to couple these equations, in a manner which is convergent, uses the following notation to describe terms in the momentum equation, *e.g.* Eq. (2.67), *excluding* $\nabla p$

$$A\mathbf{u} - \mathbf{H}(\mathbf{u}) \equiv \frac{\partial \mathbf{u}}{\partial t} + \nabla \bullet (\mathbf{u}\mathbf{u}) - \nabla \bullet (\nu \nabla \mathbf{u}) - \frac{\rho(T,p)}{\rho_0}\mathbf{g}, \tag{5.16}$$

where:

- $A\mathbf{u}$ is a linear term in $\mathbf{u}$;
- $-\mathbf{H}(\mathbf{u})$ is a function of $\mathbf{u}$ and other sources.

### Momentum corrector

A *momentum corrector equation* is formed by expressing the momentum equation, *e.g.* Eq. (2.67) in terms of the notation of Eq. (5.16), and rearranging to give

latest $\mathbf{u}$ and $p$

$$\mathbf{u} := \frac{\mathbf{H}(\mathbf{u})}{A} - \frac{1}{A}\nabla p. \tag{5.17}$$

The equation provides an update to $\mathbf{u}$, based on current values of $\mathbf{u}$ and $p$ substituted on the r.h.s. In other words, $\mathbf{H}(\mathbf{u})$ and $\nabla p$ are calculated explicitly.

In the algorithms described in the following sections, $A$ is chosen to be the extracted diagonal coefficients $a_{i,i}$ $i = 1, 2, \ldots, N$ of the matrix $\mathbf{A}$ corresponding to the momentum terms in Eq. (5.16).

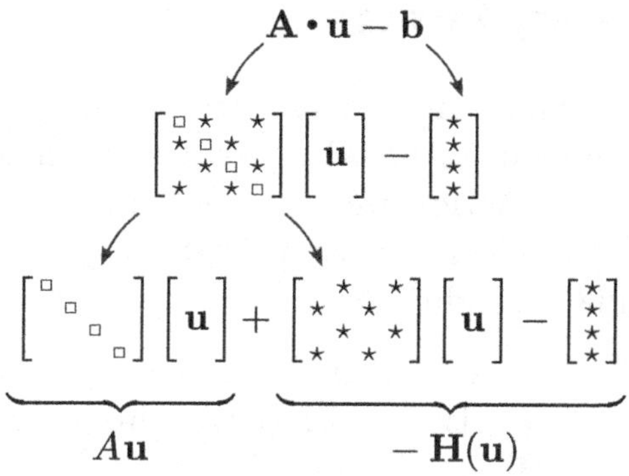


The split between $A\mathbf{u}$ and $\mathbf{H}(\mathbf{u})$ is shown above. As a matrix with diagonal components only, $A$ has one value per cell so can be represented as a scalar field. Setting $A$ to be the diagonal coefficients is a natural choice for implicit treatment of $\mathbf{u}$ within the coupling algorithm.

## Flux corrector

A *flux corrector equation* follows from Eq. (5.17) by interpolating $\mathbf{u}$ to cell faces and evaluating $\phi_\mathrm{f} = \mathbf{S}_\mathrm{f} \bullet \mathbf{u}_\mathrm{f}$ according to

$$\phi_\mathrm{f} := \mathbf{S}_\mathrm{f} \bullet \left( \frac{\mathbf{H}(\mathbf{u})}{A} \right)_\mathrm{f} - \left( \frac{|\mathbf{S}_\mathrm{f}|}{A} \right)_\mathrm{f} \nabla_\mathbf{n}\, p_\mathrm{f} \,. \tag{5.18}$$

## Pressure equation

A *pressure equation* is then created by substituting fluxes $\phi_\mathrm{f}$ from Eq. (5.18) into the mass conservation Eq. (2.46) in discrete form $\nabla \bullet \mathbf{u} \approx \sum_\mathrm{f} \phi_\mathrm{f} = 0$. The resulting expression is equivalent to a discretised pressure equation, with coefficients containing $A$ and $\mathbf{H}(\mathbf{u})$,

$$\nabla \bullet \frac{1}{A} \nabla p = \nabla \bullet \left[ \frac{\mathbf{H}(\mathbf{u})}{A} \right] . \tag{5.19}$$

## 5.11 Boundary fluxes

The equations in Sec. 5.10 are used to compute the velocity $\mathbf{u}$, flux $\phi_{\mathrm{f}}$ and pressure $p$ based on conservation of mass and momentum. At boundaries, the flux corrector Eq. (5.18) must calculate $\phi_{\mathrm{f}}$ in a manner consistent with $\mathbf{u}$ and $p$ and their respective boundary conditions.

For example, at an impermeable stationary wall the calculated flux must be $\phi_{\mathrm{f}} = 0$, consistent with the no-slip condition $\mathbf{u} = \mathbf{0}$. At boundary faces, $(\mathbf{H}(\mathbf{u})/A)_{\mathrm{f}}$ and $\nabla_{\mathbf{n}} p$ must be compatible to evaluate the correct $\phi_{\mathrm{f}}$ according to Eq. (5.18).

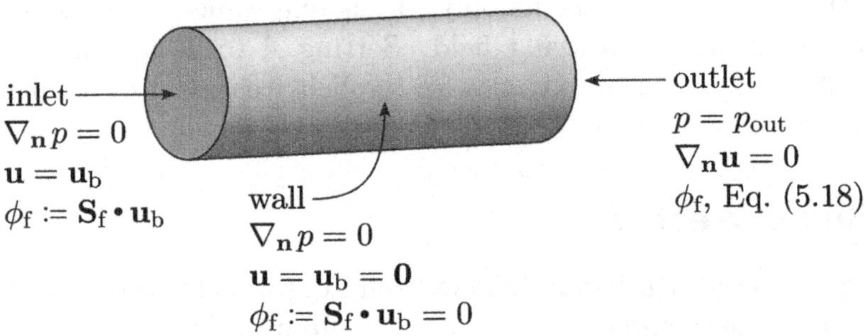


The figure shows the fundamental boundary conditions from Sec. 4.3 and corresponding flux evaluations. At the inlet and wall boundaries, $\phi_{\mathrm{f}}$ is directly assigned from the boundary velocity $\mathbf{u}_{\mathrm{b}}$ by $\phi_{\mathrm{f}} := \mathbf{S}_{\mathrm{f}} \bullet \mathbf{u}_{\mathrm{b}}$, within the flux corrector Eq. (5.18).

In the absence of body forces, the $\nabla_{\mathbf{n}} p = 0$ boundary condition is commonly applied, as discussed in Sec. 4.4. The flux $\phi_{\mathrm{f}}$ is then equivalent to assigning $(\mathbf{H}(\mathbf{u})/A)_{\mathrm{f}} := \mathbf{u}_{\mathrm{b}}$ in Eq. (5.18).

At the outlet, $\phi_{\mathrm{f}}$ is not prescribed since $\mathbf{u}$ is not a fixed value condition. Instead, $\phi_{\mathrm{f}}$ is evaluated from Eq. (5.18) using $(\mathbf{H}(\mathbf{u})/A)_{\mathrm{f}}$ taken from cells adjacent to the boundary and $\nabla_{\mathbf{n}} p_{\mathrm{f}}$ calculated on the boundary.

## Fluxes with a body force

When a body force $\mathbf{b}$ is present in the momentum equation, the gradient condition for $p$, *e.g.* at an inlet or wall, in principle becomes $\nabla_{\mathbf{n}} p = \mathbf{n} \bullet \mathbf{b}$, as discussed in Sec. 4.4. The precise details of the boundary condition in fact depend on how $\mathbf{b}$ is incorporated within the coupling algorithm, discussed below.

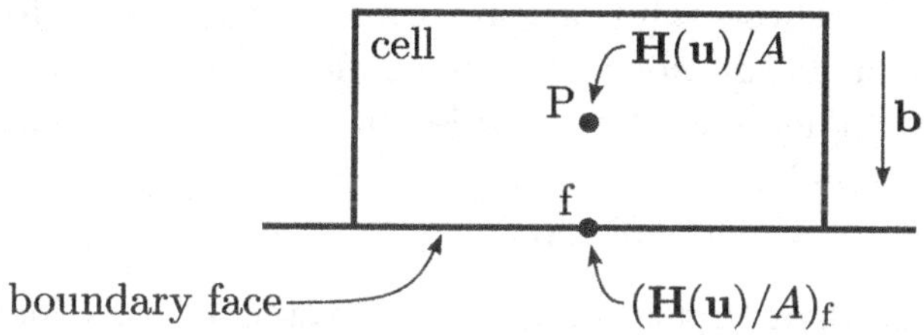


One approach, illustrated by the algorithm in Sec. 5.10, is to include the body force $\mathbf{b} = (\rho/\rho_0)\mathbf{g}$ within $\mathbf{H}(\mathbf{u})$ in Eq. (5.16). In that case, the assignment $(\mathbf{H}(\mathbf{u})/A)_\mathrm{f} := \mathbf{u}_\mathrm{b}$ cannot be valid, so $(\mathbf{H}(\mathbf{u})/A)_\mathrm{f} := \mathbf{H}(\mathbf{u})/A$ is adopted instead.

With $(\mathbf{H}(\mathbf{u})/A)_\mathrm{f}$ established, the $\nabla_{\mathbf{n}} p$ condition is calculated based on the known $\phi_\mathrm{f}$ by inverting Eq. (5.18). This approach causes $\nabla_{\mathbf{n}} p$ to include a contribution from viscous stresses, as in Eq. (4.5), which may cause instability.

To avoid this problem, $\mathbf{b}$ is omitted from $\mathbf{H}(\mathbf{u})$ in Eq. (5.16), appearing instead as an extra term in the other equations in Sec. 5.10, *e.g.* the flux corrector Eq. (5.18) which becomes

$$\phi_\mathrm{f} := \mathbf{S}_\mathrm{f} \bullet \left(\frac{\mathbf{H}(\mathbf{u})}{A}\right)_\mathrm{f} - \left(\frac{|\mathbf{S}_\mathrm{f}|}{A}\right)_\mathrm{f} \nabla_{\mathbf{n}} p_\mathrm{f} + \mathbf{S}_\mathrm{f} \bullet \left(\frac{\mathbf{b}}{A}\right)_\mathrm{f}. \qquad (5.20)$$

Assigning $(\mathbf{H}(\mathbf{u})/A)_\mathrm{f} := \mathbf{u}_\mathrm{b}$ at boundaries where $\mathbf{u}$ is fixed satisfies Eq. (5.20) when $\nabla_{\mathbf{n}} p = (\mathbf{S}_\mathrm{f}/|\mathbf{S}_\mathrm{f}|) \bullet \mathbf{b} = \mathbf{n} \bullet \mathbf{b}$. This is the condition ignoring viscous stresses described in Sec. 4.4.

## 5.12 Steady-state solution

The equations in Sec. 5.10 are combined using algorithms that couple the solutions for $p$ and $\mathbf{u}$. One algorithm is SIMPLE (Semi-Implicit Method for Pressure-Linked Equations)†, which is presented here with a modern interpretation.

The SIMPLE algorithm is generally used to produce steady flow solutions in CFD. These solutions are directly applicable for flows that reach a *steady state*, *i.e.* when flow variables stop changing in time. They can also provide approximate solutions to flows that are moderately unsteady, usually at a lower cost than a more exact transient solution.

An example of the algorithm is shown for the system of equations presented in Sec. 5.9. The time derivative ($\partial/\partial t$) terms are omitted due to the steady-state assumption.

The algorithm involves an iterative sequence with steps, $i = 1, 2, \ldots, i_{\text{end}}$. It begins by constructing a matrix equation for energy which is under-relaxed by a factor $\alpha_T$. The equation is solved for $T$, which is used to update $\rho(T, p)$ according to an equation of state. A matrix equation is then constructed using all the terms from the momentum equation *excluding* $\nabla p$, *i.e.*

$$\mathbf{A} \bullet \mathbf{u} - \mathbf{b} \equiv \nabla \bullet (\mathbf{u}\mathbf{u}) - \nabla \bullet (\nu \nabla \mathbf{u}) - \frac{\rho(T, p)}{\rho_0}\mathbf{g} \qquad (5.21)$$

The matrix equation is under-relaxed by a factor $\alpha_{\mathbf{u}}$ before equating with $-\nabla p$ and solving for $\mathbf{u}$ (the *momentum predictor*).

$A\mathbf{u}$ and $\mathbf{H}(\mathbf{u})$ are then evaluated from $\mathbf{A} \bullet \mathbf{u} - \mathbf{b}$ (the *momentum matrix*), as described on page 166. They are used to form the *pressure equation*, which is solved for $p$.

---

†Suhas Patankar and Brian Spalding, *A calculation procedure for heat, mass and momentum transfer in three-dimensional parabolic flows*, 1972.

The new pressure $p$ is used to correct the flux $\phi_\mathrm{f}$ so that it obeys mass conservation better (the *flux corrector*). It is then under-relaxed by a factor $\alpha_p$ before correcting $\mathbf{u}$ before the next solution step begins (the *momentum corrector*).

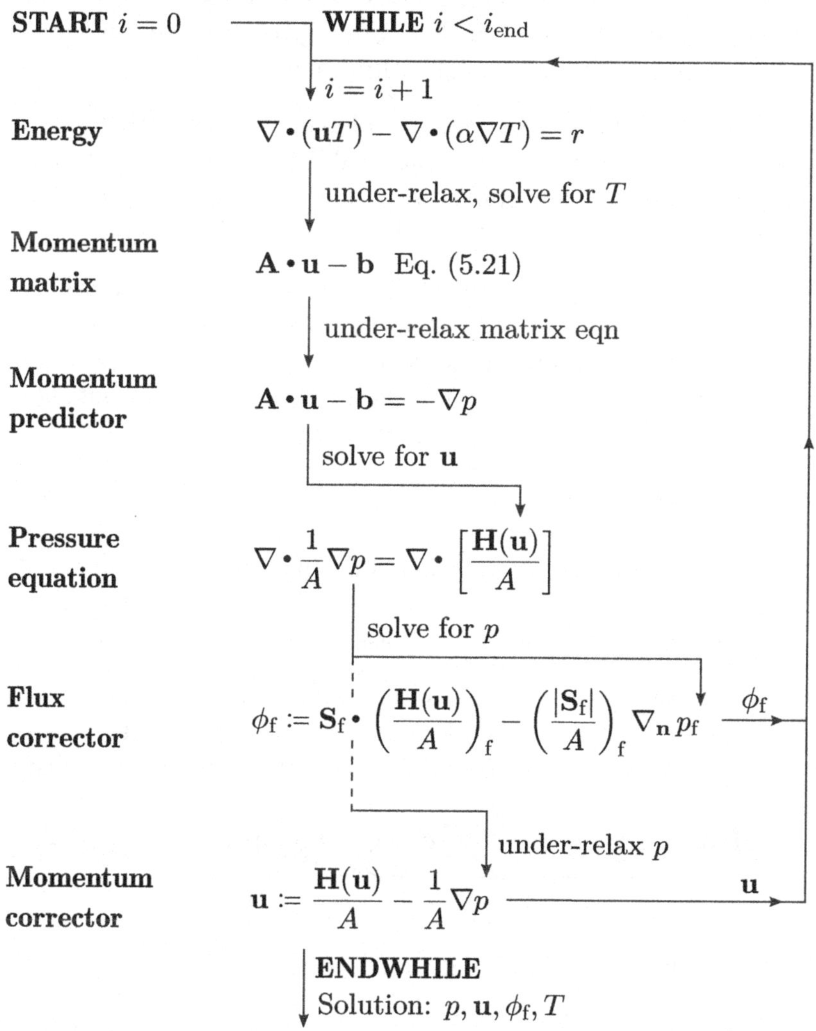

## 5.13 Steady-state convergence

The absence of a time derivative from an equation written in steady-state form reduces the diagonal dominance, and hence convergence, of the resulting matrix equation, as discussed in Sec. 5.5. Under-relaxation is therefore applied to $T$ $\mathbf{u}$, and $p$ to promote convergence in the algorithm in Sec. 5.12.

The $T$ and $\mathbf{u}$ fields use equation under-relaxation with factors $\alpha_T$ and $\alpha_\mathbf{u}$, respectively. A value of 0.7 is commonly applied, decreasing to 0.5 for less convergent cases (and sometimes to 0.3 in compressible flow cases, beyond the scope of this book).

Under-relaxation of $p$ is more subtle. The flux corrector requires that $p$ is not under-relaxed to ensure $\phi_\mathrm{f}$ obeys mass conservation better. For the momentum corrector, field under-relaxation is subsequently applied to $p$ with a factor $\alpha_p$. To find an optimal $\alpha_p$ for the momentum corrector, we examine

$$A\mathbf{u} = \mathbf{H}(\mathbf{u}) - \nabla p. \tag{5.22}$$

This explicit momentum equation contains diagonal coefficients $A$ and an off-diagonal contribution $\mathbf{H}(\mathbf{u}) = -\sum a_\mathrm{N}\mathbf{u}_\mathrm{N}$ from neighbour cell coefficients $a_\mathrm{N}$ and associated velocities $\mathbf{u}_\mathrm{N}$.

Convergence is compromised by the explicit nature of $\mathbf{H}(\mathbf{u})$. It can be more implicit by "adding and subtracting" $A^\star\mathbf{u}$, where coefficients $A^\star = -\sum a_\mathrm{N}$ are applied to $\mathbf{u}$ in the "owner" cells:

$$\mathbf{H}(\mathbf{u}) = \underbrace{\left(\sum a_\mathrm{N}\right)\mathbf{u} - \sum a_\mathrm{N}\mathbf{u}_\mathrm{N}}_{\mathbf{H}^\star(\mathbf{u})} \underbrace{- \left(\sum a_\mathrm{N}\right)\mathbf{u}}_{+A^\star\mathbf{u}}. \tag{5.23}$$

Combining Eq. (5.22) and Eq. (5.23) gives

$$(A - A^\star)\mathbf{u} = \mathbf{H}^\star(\mathbf{u}) - \nabla p, \tag{5.24}$$

which attempts to make Eq. (5.22) more implicit in $\mathbf{u}$.[†] The pressure equation derived from Eq. (5.24) is:

$$\nabla \bullet \frac{1}{A - A^\star} \nabla p = \nabla \bullet \left[ \frac{\mathbf{H}^\star(\mathbf{u})}{A - A^\star} \right] \tag{5.25}$$

This corresponds to under-relaxation of Eq. (5.19) by $\alpha_p$, where $\alpha_p A = A - A^\star$. A momentum matrix with coefficients $A^\star$ is approximately diagonally equal since there is no time derivative. Since $A$ represents the diagonal coefficients under-relaxed by $\alpha_\mathbf{u}$, $A^\star = \alpha_\mathbf{u} A$. Relating the expressions for $A$ and $A^\star$ gives the optimal under-relaxation factor for $p$ for convergence as

$$\alpha_p = 1 - \alpha_\mathbf{u}, \tag{5.26}$$

leading to the popular choice of $\alpha_\mathbf{u} = 0.7$ and $\alpha_p = 0.3$.

## Residual control

The algorithm in Sec. 5.12 uses a fixed number of solution steps $i_\text{end}$. In practice, $i_\text{end}$ must be chosen to be large enough to reach an acceptable level of convergence. Once convergence is reached, the simulation should stop to avoid unnecessary computing cost.

A common stopping criterion applies a residual level for each equation, below which the equation is deemed to be converged. When all equations satisfy their respective residual controls, the simulation then stops.

Convergence can also be determined by monitoring any suitable metric, including objective measurements from the simulation, *e.g.* a force coefficient. When the metric no longer changes significantly over subsequent steps, the simulation is stopped.

---

[†] Jeff Van Doormaal and George Raithby, *Enhancements of the SIMPLE method for predicting incompressible fluid flows*, 1984.

## 5.14 Descent methods

The Gauss-Seidel method, introduced in Sec. 5.2, provides convergent solutions for many problems in CFD. It is most effective when a modest reduction in residual is required, *e.g.* as part of a steady-state solution described in Sec. 5.12.

When the Gauss-Seidel method requires a lot of sweeps (*e.g.* over 10) to converge to a suitable tolerance, alternative methods may be more efficient. *Descent methods* provide alternative matrix solvers that are often used in CFD.

Descent methods represent the equations which are being solved as a *minimisation* problem. It is demonstrated below using a matrix equation of the form $\mathbf{A} \bullet \mathbf{\Psi} = \mathbf{b}$ with 2 values

$$\begin{bmatrix} 6 & -2 \\ -2 & 4 \end{bmatrix} \begin{bmatrix} \mathbf{\Psi}_1 \\ \mathbf{\Psi}_2 \end{bmatrix} = \begin{bmatrix} 1 \\ 3 \end{bmatrix}. \tag{5.27}$$

The minimisation presents the equation in *quadratic form*; in matrix notation, it is a *scalar* function of the form

$$f(\mathbf{\Psi}) = \frac{1}{2}\mathbf{\Psi} \bullet \mathbf{A} \bullet \mathbf{\Psi} - \mathbf{b} \bullet \mathbf{\Psi} \tag{5.28}$$

The quadratic form of Eq. (5.27), is:

$$f(\mathbf{\Psi}) = 3\mathbf{\Psi}_1^2 + 2\mathbf{\Psi}_2^2 - 2\mathbf{\Psi}_1\mathbf{\Psi}_2 - \mathbf{\Psi}_1 - 3\mathbf{\Psi}_2 \tag{5.29}$$

The gradient of $f(\mathbf{\Psi})$ is:

$$\frac{\partial f(\mathbf{\Psi})}{\partial \mathbf{\Psi}} = \begin{bmatrix} \partial f/\partial \mathbf{\Psi}_1 \\ \partial f/\partial \mathbf{\Psi}_2 \end{bmatrix} = \begin{bmatrix} 6\mathbf{\Psi}_1 - 2\mathbf{\Psi}_2 - 1 \\ -2\mathbf{\Psi}_1 + 4\mathbf{\Psi}_2 - 3 \end{bmatrix} \tag{5.30}$$

Critically, $\partial f(\mathbf{\Psi})/\partial \mathbf{\Psi} = \mathbf{A} \bullet \mathbf{\Psi} - \mathbf{b} = -\mathbf{r}$, *i.e.* the negative of the residual vector Eq. (5.10), as verified by the model example.

Equating the gradient to zero, $\partial f(\boldsymbol{\Psi})/\partial\boldsymbol{\Psi} = \mathbf{0}$, corresponds to a *minimum in the quadratic function.* At the same time, it is the *solution* to $\mathbf{A} \bullet \boldsymbol{\Psi} = \mathbf{b}$. The method is therefore concerned with finding the minimum of the quadratic form efficiently.

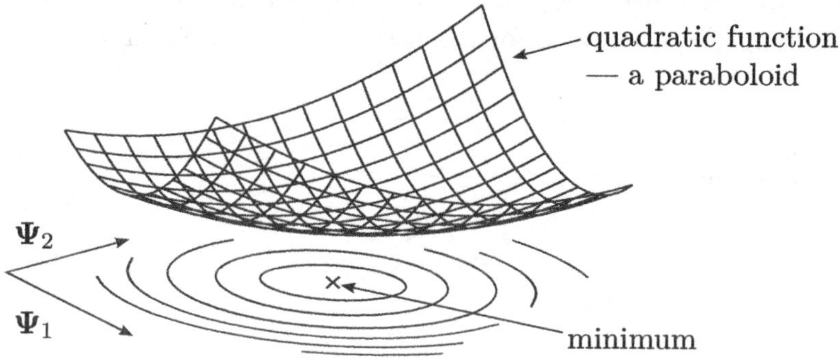


For this method to work, the quadratic form must have a minimum, which requires that $\mathbf{A}$ is *symmetric* and *positive-definite.* A positive-definite matrix is hard to visualise, but for a 2-value function it ensures the quadratic function is a paraboloid.

Diagonal dominance is the convergence condition for the Gauss-Seidel method, discussed in Sec. 5.3. Importantly, a symmetric matrix that is diagonally dominant, and has positive diagonal coefficients, is also positive-definite.

## Matrix operations

The '$\bullet$' operation between two single-column matrices, *e.g.* $\mathbf{b} \bullet \boldsymbol{\Psi}$, in Eq. (5.28) is represented in other texts using matrix notation by a transpose $[\mathbf{b}]^{\mathrm{T}}[\boldsymbol{\Psi}]$. For the example in Eq. (5.27), it is:

$$\mathbf{b} \bullet \boldsymbol{\Psi} = [\mathbf{b}]^{\mathrm{T}}[\boldsymbol{\Psi}] = \begin{bmatrix} 1 & 3 \end{bmatrix} \begin{bmatrix} \Psi_1 \\ \Psi_2 \end{bmatrix} = \Psi_1 + 3\Psi_2 \quad (5.31)$$

## 5.15 Conjugate gradient method

The solution to a matrix equation $\mathbf{A} \bullet \boldsymbol{\Psi} = \mathbf{b}$ by descent methods involves finding the minimum of the equation in quadratic form. This can be illustrated using the contours of the paraboloid describing the quadratic for the case with values $\boldsymbol{\Psi}_1$ and $\boldsymbol{\Psi}_2$.

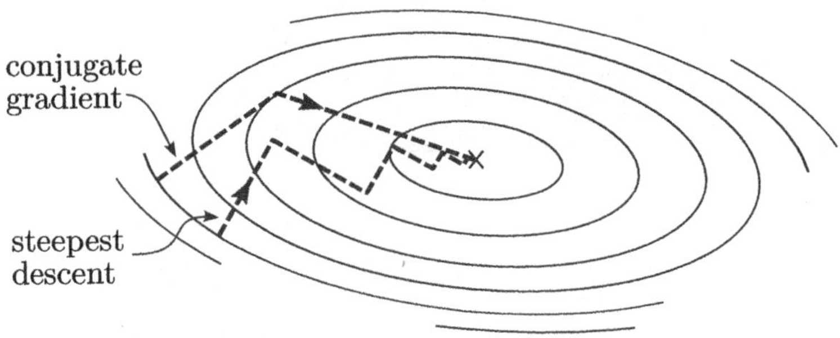


The search for the minimum involves a series of updates to $\boldsymbol{\Psi}$ of the form

$$\boldsymbol{\Psi} := \boldsymbol{\Psi}^{\mathrm{c}} + \alpha \mathbf{p} \tag{5.32}$$

where $\boldsymbol{\Psi}$ is the next update using latest (current) values $\boldsymbol{\Psi}^{\mathrm{c}}$. The column vector $\mathbf{p}$ provides the *direction* of the line search towards the minimum; the scalar $\alpha$ provides the magnitude of the line is that direction.

### Steepest descent

The intuitive way to reach the minimum is to follow the direction of *steepest descent.* The method is fairly simple because the direction is defined by the negative of gradient of the quadratic form $\partial f(\boldsymbol{\Psi})/\partial \boldsymbol{\Psi}$ which is the residual $\mathbf{r} = \mathbf{b} - \mathbf{A} \bullet \boldsymbol{\Psi}$.

The distance to "walk" is naturally until the lowest point is reached, corresponding to[†]

$$\alpha = \frac{\mathbf{r} \bullet \mathbf{r}}{\mathbf{r} \bullet \mathbf{A} \bullet \mathbf{r}}. \tag{5.33}$$

The method reaches the minimum point along a zigzag path where consecutive directions are orthogonal. The directions do not change and result in a *larger* number of very short steps, the *more* the paraboloid is stretched in one direction.

## The conjugate direction

The **conjugate gradient** (CG) method chooses search directions that are *conjugate* with **A**. This means each new direction $\mathbf{p}$ corresponds to the previous one $\mathbf{p}^{\mathrm{c}}$, satisfying

$$\mathbf{p}^{\mathrm{c}} \bullet \mathbf{A} \bullet \mathbf{p} = 0. \tag{5.34}$$

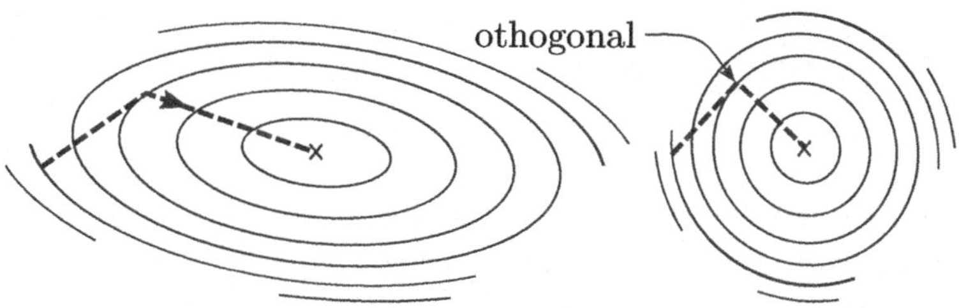


This can be imagined as the directions being orthogonal *with the stretch in the paraboloid removed.* For 2 values, CG finds the minimum in 2 steps, rather than several zigzag steps.

CG provides the basis for practical matrix solvers for CFD, described in Sec. 5.16. For a detailed explanation of the CG method, see the reference below.

---

[†]Jonathan Richard Shewchuk *An introduction to the conjugate gradient method without the agonizing pain*, 1994.

## 5.16 Preconditioning and asymmetry

The conjugate gradient (CG) method provides a good basis for a matrix solver in CFD. In theory, it provides an exact solution in $N$ steps for a problem with $N$ values. In practice, there is an accumulation of rounding errors from finite precision arithmetic, such that an exact solution is not attained. Problems are also too large for it to be feasible to run $N$ steps.

CG is instead used as an iterative method, aiming to get the solution residual within an acceptable tolerance. The rate of convergence is therefore critical, which can be shown to be slower for increasing *condition number* of the matrix, $\kappa = \lambda_{\text{max}}/\lambda_{\text{min}}$, where $\lambda_{\text{max}}$ and $\lambda_{\text{min}}$ are its largest *eigenvalues.*

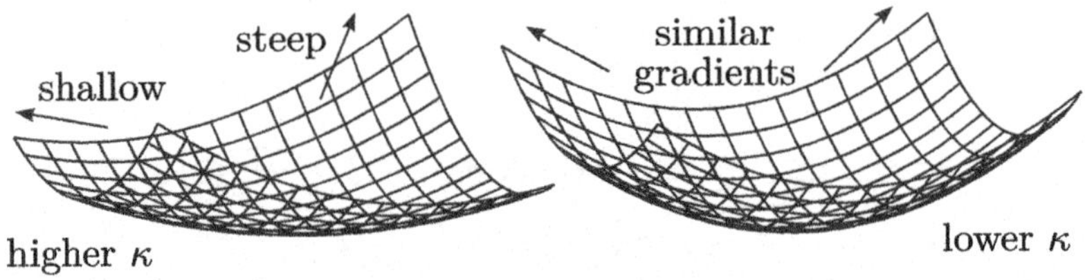


In the example with two values, the eigenvalues correspond to the gradients of the principal axes of the paraboloid described by the quadratic function. Lower $\kappa$ corresponds to closer gradients, *i.e.* to a rounder "bowl" shape.

The best convergence corresponds to $\kappa = 1$ when $\mathbf{A} = \mathbf{I}$ ($\mathbf{I}$ is the identity matrix). *Preconditioning* finds a matrix $\mathbf{M}$, similar to $\mathbf{A}$, which can be inverted and multiplied to the equation as follows:

$$(\mathbf{M}^{-1} \bullet \mathbf{A}) \bullet \boldsymbol{\Psi} = \mathbf{M}^{-1} \bullet \mathbf{b}. \tag{5.35}$$

The aim of preconditioning is to improve convergence by choosing a preconditioner $\mathbf{M}^{-1}$ so that $\mathbf{M}^{-1} \bullet \mathbf{A}$ is closer to $\mathbf{I}$ than $\mathbf{A}$.

Effective methods for calculating the preconditioner $\mathbf{M}^{-1}$ are:

- diagonal-based incomplete Cholesky (DIC) for symmetric $\mathbf{A}$;
- diagonal-based incomplete LU (DILU) for an asymmetric $\mathbf{A}$.

## Biconjugate gradient stabilised method

The **preconditioned CG** (PCG) method minimises the quadratic form which requires that $\mathbf{A}$ is symmetric. However, any equation containing an advection term produces an asymmetric matrix (see Sec. 5.1), when PCG is unsuitable.

The **preconditioned biconjugate gradient** (PBiCG) method is designed for asymmetric matrices. It augments the calculations of residuals and line search directions in the CG method with a second set of calculations relating to the transpose matrix $\mathbf{A}^{\mathrm{T}}$. The computational cost is approximately $2\times$ the cost of PCG.

The convergence behaviour of PBiCG is somewhat erratic, with large variations in the reduction of the residual between successive iterations. Sometimes the method breaks down and fails.

The **preconditioned biconjugate gradient stabilised** method (PBiCGStab) is a modification of PBiCG. It is exhibits much smoother convergence and robustness than PBiCG so it is far more preferable for CFD applications.

Recommended CG-based methods are summarised below.

- *symmetric*: **PCG** with **DIC** preconditioning.
- *asymmetric*: **PBiCGStab** with **DILU** preconditioning.

The implementation of preconditioning and CG-based methods can be found in: Barrett R. *et al. Templates for the solution of linear systems*, available at `http://www.netlib.org`.

## 5.17 Multi-grid method

The methods for solving matrix equations described so far are Gauss-Seidel and CG. They are iterative, and require few iterations to calculate changes in the field, when the change at any given point is influenced only by points in the close vicinity.

This makes them efficient for solving equations whose range of influence is limited, *e.g.* by the flow speed through advection $\nabla \bullet (\mathbf{u}\Psi)$ in a transport equation.

The pressure equation contains neither a local rate of change $\partial/\partial t$ nor advection term (assuming $\rho$ = constant). A disturbance at any point influences the solution everywhere in the domain instantaneously, as discussed in Sec. 4.3.

To transfer changes across a domain, Gauss-Seidel might require as many sweeps as the number of cells across a domain, which would be impractical. Some method is therefore needed that transfers change across the domain more efficiently.

The multi-grid method† uses a coarse mesh to overcome the problem of slow transfer of change across the domain, reducing the error at "low-frequency". It transfers that solution through a succession of finer meshes to be accurate at higher frequencies.

The multi-grid method works by first calculating the matrix $\mathbf{A}$ and residual vector $\mathbf{r}$ on the original (finest) mesh. Coarser meshes are then successively formed until the mesh reaches its coarsest level, containing as few as 10 cells.

The matrix $\mathbf{A}$ and $\mathbf{r}$ are formed at each level of coarsening. Different methods exist to calculate $\mathbf{A}$ and $\mathbf{r}$, including a simple *agglomeration* discussed in Sec. 5.18.

---

†Achi Brandt, *Multi-level adaptive solutions to boundary-value problems*, 1977.

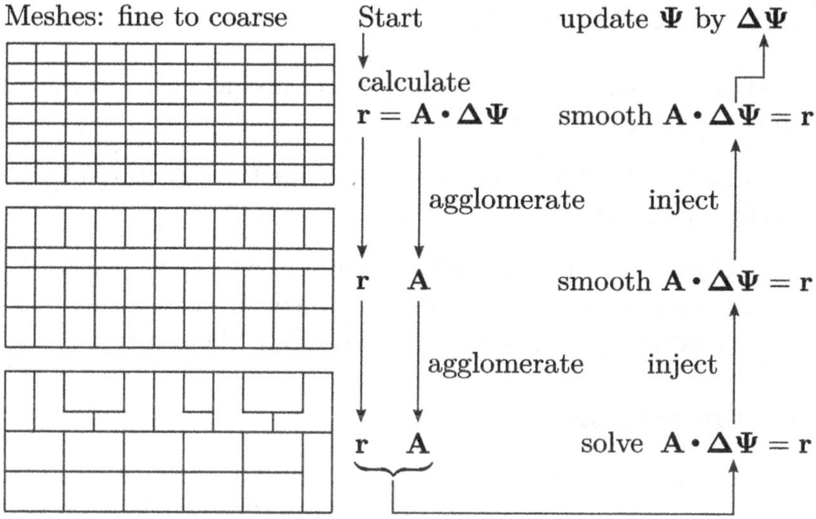


Equations are constructed in terms of the *correction* $\mathbf{\Delta\Psi}$, required to make $\mathbf{\Psi}$ exact, related to the residual $\mathbf{r}$ by

$$\begin{aligned}\mathbf{r} &= \mathbf{b} - \mathbf{A} \bullet \mathbf{\Psi} = \mathbf{A} \bullet \mathbf{\Psi}_{\text{ex}} - \mathbf{A} \bullet \mathbf{\Psi} \\ &= \mathbf{A} \bullet \mathbf{\Delta\Psi}.\end{aligned} \tag{5.36}$$

At the coarsest level, $\mathbf{A} \bullet \mathbf{\Delta\Psi} = \mathbf{r}$ is solved precisely for $\mathbf{\Delta\Psi}$, which can be performed efficiently using a direct solver since the matrix is small.

$\mathbf{\Delta\Psi}$ is then *injected* into the finer level, providing the initial $\mathbf{\Delta\Psi}$ for a few sweeps of the Gauss-Seidel method at that level. Smoothing and injection are repeated up to the finest level, when the final $\mathbf{\Delta\Psi}$ is applied to $\mathbf{\Psi}$ to yield the solution.

It is generally more efficient to do more Gauss-Seidel sweeps at the coarser levels, when the cost if low due to a smaller matrix size. For example, 4 sweeps may be applied at the coarser levels, while 2 sweeps are applied at the finest.

## 5.18 GAMG method

**GAMG** is an effective form of multi-grid method used in finite volume CFD. It combines:

- geometric ("G") agglomeration to define the structure of the coarse meshes;
- the *algebraic* multi-grid (AMG) method, where the matrix **A** is constructed at a coarser level from coefficients at the finer level, rather than by geometric data from the coarse mesh.

### Agglomeration

*Pairwise agglomeration* forms coarser meshes by joining pairs of cells at each level of coarsening. In a sweep through the cells, an unpaired cell is paired with the (unpaired) neighbour that shares the face with largest area. This method generally maintains a low aspect ratio (see Sec. 8.2) in the resulting agglomerated cells.

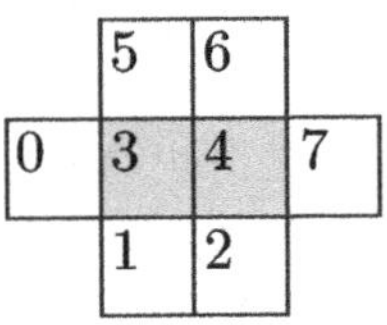


cell: 0 1 2 3 4 5 6 7

$$\begin{bmatrix} 1 & 1 & 0 & -4 & 1 & 1 & 0 & 0 \\ 0 & 0 & 1 & 1 & -4 & 0 & 1 & 1 \end{bmatrix} \begin{bmatrix} \Delta\boldsymbol{\Psi}_3 \\ \Delta\boldsymbol{\Psi}_4 \end{bmatrix} = \begin{bmatrix} 3 \\ 2 \end{bmatrix}$$

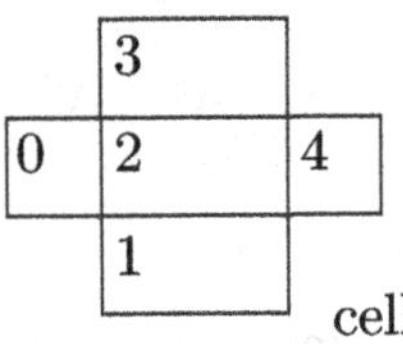


$$\begin{bmatrix} 1 & 2 & -6 & 2 & 1 \end{bmatrix} \begin{bmatrix} \Delta\boldsymbol{\Psi}_2 \end{bmatrix} = \begin{bmatrix} 5 \end{bmatrix}$$

cell: 0 1 2 3 4

The example shows an agglomeration of cell pairs 1-2, 3-4 and 5-6, which form cells 1, 2 and 3 respectively in the coarse mesh. If we solve the equation $\nabla \cdot (\Gamma \nabla \Psi) = f$ on a 2D mesh of square cells with $|\mathbf{S}_f| = |\boldsymbol{\Delta}\mathbf{d}| = \Gamma = 1$, then Eq. (3.2) and Eq. (3.5) calculate diagonal coefficients $a_{ii} = -4$ and non-zero off-diagonal coefficients $a_{ij} = 1$.

**Algebraic multi-grid** creates coefficients in the coarser mesh by summing coefficients $\mathbf{A}$ and source $\mathbf{r}$ from the finer mesh. The example produces one row of $\mathbf{A}$ which is $\begin{bmatrix} 1 & 2 & -6 & 2 & 1 \end{bmatrix}$.

If the Laplacian for the coarse mesh were discretised directly, the coefficients in that row would be $\begin{bmatrix} 2/3 & 2 & -5\,1/3 & 2 & 2/3 \end{bmatrix}$ due to increasing $|\boldsymbol{\Delta}\mathbf{d}|$. This discrepancy between agglomerated and calculated coefficients is repeated in subsequent agglomerations.

Multi-grid begins solving/smoothing for $\boldsymbol{\Delta\Psi}$ (with $\mathbf{A} \bullet \boldsymbol{\Delta\Psi} = \mathbf{r}$) at the coarsest mesh. Coarse cell values of $\boldsymbol{\Delta\Psi}$ are then injected into corresponding cells in the next finest mesh.

With a Laplacian term, the coefficient discrepancy caused by algebraic agglomeration causes $\boldsymbol{\Delta\Psi}$ to be under-predicted. A correction to $\boldsymbol{\Delta\Psi}$ can be applied by scaling it by a factor

$$\alpha = \frac{\mathbf{r} \bullet \boldsymbol{\Delta\Psi}}{\boldsymbol{\Delta\Psi} \bullet \mathbf{A} \bullet \boldsymbol{\Delta\Psi}}. \tag{5.37}$$

This scaling is used when the equation being solved is dominated by a Laplacian term, *e.g.* a pressure equation, when summing coefficients between agglomerated cells is inaccurate. Scaling is not generally used for transport dominated by advection, since there is no error associated with summing coefficients from the advective derivative.

Equation 5.37 is derived by minimising the error with respect to $\alpha$ in the equation $\mathbf{A} \bullet \alpha\boldsymbol{\Delta\Psi} = \mathbf{r}$. The minimisation is performed by setting $\partial f / \partial \alpha = 0$ for the quadratic form

$$f(\boldsymbol{\Delta\Psi}) = \frac{\alpha^2}{2} \boldsymbol{\Delta\Psi} \bullet \mathbf{A} \bullet \boldsymbol{\Delta\Psi} - \alpha \mathbf{r} \bullet \boldsymbol{\Delta\Psi}. \tag{5.38}$$

## 5.19 Transient solution

A modern interpretation of the SIMPLE algorithm was described in Sec. 5.12 to couple steady solutions for $p$ and $\mathbf{u}$. An equivalent *transient* solution follows an iterative sequence in which equations for $p$ and $\mathbf{u}$ are solved over successive time intervals $\Delta t$, between a start time $t_{\text{start}}$ and end time $t_{\text{end}}$.

In a transient simulation, $\Delta t$ needs to be relatively small to maintain sufficient accuracy as the solution evolves over time $t$. The equations do not then generally require under-relaxation to converge due to the increase of $V/\Delta t$ to the diagonal coefficients of the matrix from discretisation of the time derivative $\partial/\partial t$.

A transient simulation could follow the same sequence as the SIMPLE algorithm on page 170, but iterating over that sequence within each time step is costly. A more efficient algorithm solves the sequence only once per time step but adds an iterative loop which substitutes $\mathbf{u}$ from the momentum corrector into the $\mathbf{H}(\mathbf{u})$ term of the pressure equation.

The pressure equation is then solved a second time, followed by a second momentum corrector, in the style of the PISO algorithm.† The addition of this "PISO loop" improves the accuracy of $p$, $\mathbf{u}$ and $\phi_{\text{f}}$ within each time step. The improved $\mathbf{u}$ becomes $\mathbf{u}^{\text{o}}$ in the next time iteration, which critically increases the accuracy of the time derivative $\partial\mathbf{u}/\partial t$ in the momentum equation.

Without any iteration over the entire system of equations, the advection terms are discretised using $\phi_{\text{f}}$ from the previous time step. This "lagging" of $\phi_{\text{f}}$ does not compromise accuracy significantly, if $\Delta t$ is sufficiently small, e.g. corresponding to $\text{Co} = 1$.

A further iteration of the PISO loop can be introduced to solve a

†Pressure Implicit with Splitting of Operators, 1986

third pressure equation and momentum corrector, in particular as part of an update to non-orthogonality described in Sec. 5.20. Increasing the iterations still further is generally not beneficial.

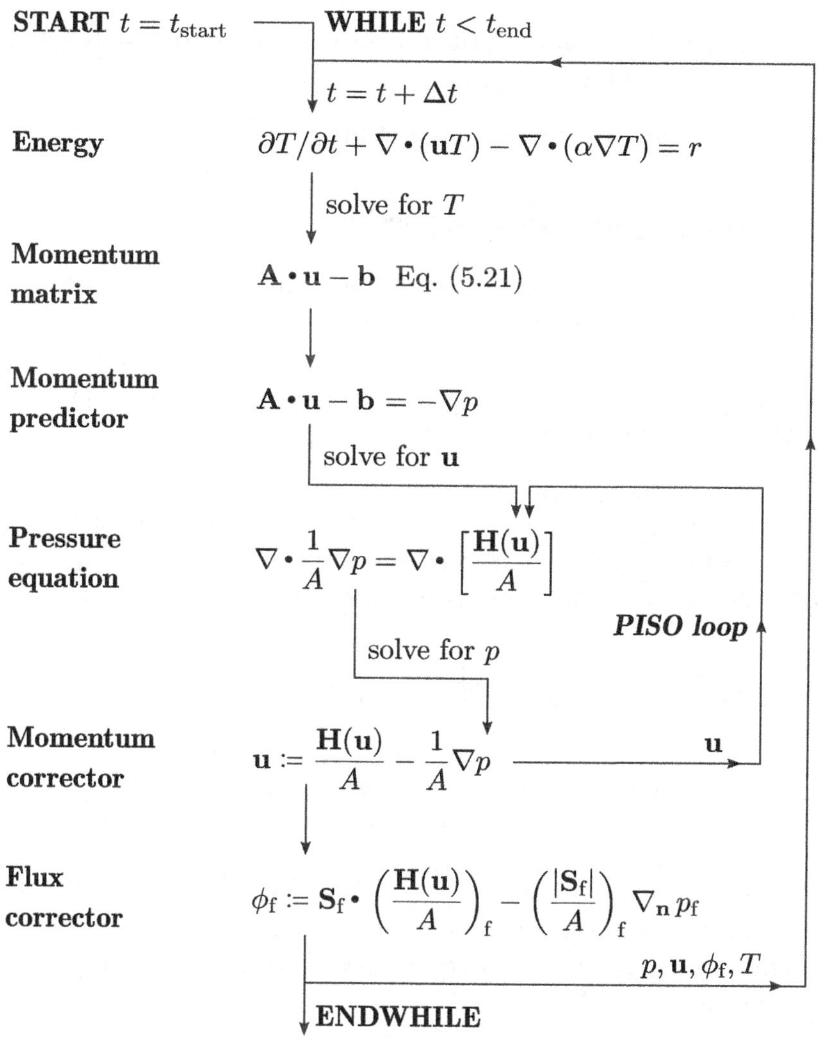

## 5.20 Transient solution controls

The following additional controls are commonly used for the transient solution described in Sec. 5.19:

- an iterative loop around the pressure equation to update any non-orthogonal correction;
- an on/off switch for the momentum predictor.

### Non-orthogonal corrector loop

The algorithm solves the pressure equation, Eq. (5.19), which includes the discretisation of the Laplacian term $\nabla \bullet (1/A)\nabla p$, according to Sec. 3.7. For a computational mesh with significant non-orthogonality, the corrected scheme Eq. (3.7) should be applied to maintain sufficient accuracy.

This correction to the pressure Laplacian is calculated using the current $p$, which produces an explicit contribution to the matrix equation. The equation can be imagined with the explicit correction as an additional term $f_{\text{corr}}(\nabla p)$.

$$\nabla \bullet \frac{1}{A}\nabla p = \nabla \bullet \left[\frac{\mathbf{H}(\mathbf{u})}{A}\right] + f_{\text{corr}}(\nabla p)$$

solve for $p$ — *non-orthogonal corrector loop*

To improve accuracy within each time step during a transient simulation, $f_{\text{corr}}(\nabla p)$ can be recalculated using the latest $p$ and the equation solved again.

The accuracy can be improved by further iterations, but this is not generally recommended due to the increased computational cost. With the pressure equation encountered twice as part of the PISO loop, and with one corrector on each occasion, the pressure matrix is solved a total of 4 times.

Solving the pressure matrix is costly, so any reduction in the number of solutions is welcome. Recognising that $f_{\text{corr}}(\nabla p)$ is updated as part of the PISO iterations anyway, an alternative solution strategy uses no non-orthogonal correction but adds an extra PISO loop, solving the pressure equation a total of 3 times.

## Solving the momentum predictor

The transient algorithm on page 184 includes the momentum predictor that provides an initial solution for $\mathbf{u}$, which is used to evaluate $\mathbf{H}(\mathbf{u})$ in the pressure equation. The solution for $p$ is then substituted into the momentum corrector which calculates a new $\mathbf{u}$.

Since $\mathbf{u}$ is ultimately recalculated by the momentum corrector — a momentum equation in explicit form — *is it necessary* to solve the momentum predictor at all? The answer is "no", unless it makes the solution more convergent which is more likely for high speed flows which are dominated more by momentum exchanges than mass conservation.

The transient algorithm should therefore include a simple switch that makes it possible to turn off the momentum predictor step. The rest of the algorithm remains the same, with the momentum matrix constructed to provide $\mathbf{H}(\mathbf{u})$ and $A$.

With this switch included, the momentum predictor is then turned off in a lot of applications, *e.g.* highly viscous flows. In practice, it is often switched off for more complex flows, *e.g.* multiphase flows, which are beyond the scope of this book.

## 5.21 The PIMPLE algorithm

The $p$-$\mathbf{u}$ coupling algorithms in Sec. 5.12 and Sec. 5.19 can be combined into an algorithm known as PIMPLE. PIMPLE merges the controls of PISO and SIMPLE (hence the merged acronym), in particular the iterative loops and under-relaxation.

All controls are optional; the standard transient algorithm is replicated by deactivating both the under-relaxation and the PIMPLE loop. By including the PIMPLE loop, equations are solved using variables updated within the time step. Accuracy is improved in particular due to the update of matrix coefficients from the contribution of $\phi_{\mathrm{f}}$ to advection.

For transient simulations, temporal accuracy can be maintained at a higher Co ($> 1$) using a second order time scheme (Sec. 3.18) combined with iterations of the PIMPLE loop. Similarly, the PIMPLE loop can update explicit source terms, *e.g.* in energy or momentum, to improve accuracy.

### Pseudo-transient solution

PIMPLE can be configured to produce a steady flow solution quickly by a *pseudo-transient* simulation. These simulations are not intended to capture realistic transient behaviour so can run at Co $\gg 1$ with some under-relaxation if necessary.

The simulations can be accelerated to a steady state using *local time stepping* (LTS). LTS recognises that $\Delta t$ is limited by the maximum Co associated with the cell with small $V$ and/or high $|\mathbf{u}|$. It uses a field of $\Delta t^{-1}$ corresponding to the Co limit *in each cell* to accelerate the transient solution. While using a $\Delta t^{-1}$ field makes the transient solution invalid, it is acceptable at steady state when $\partial/\partial t = 0$.

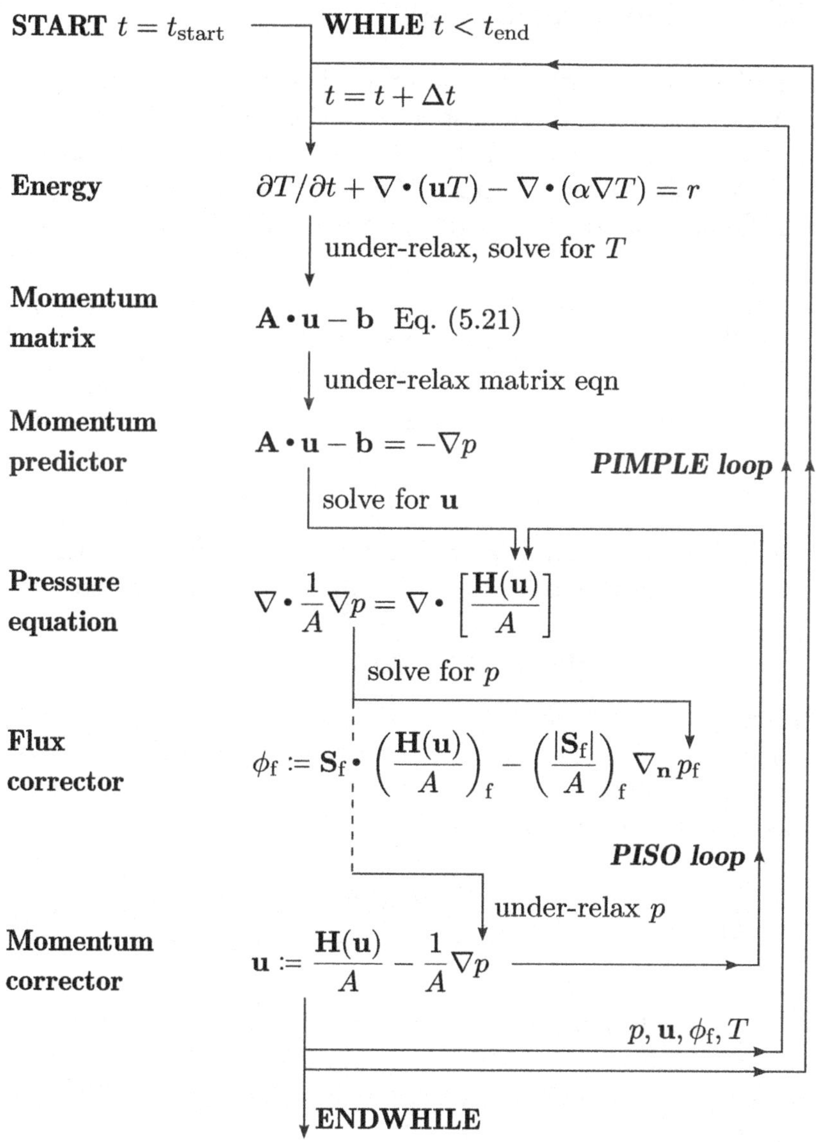
**START** $t = t_{\text{start}}$
**WHILE** $t < t_{\text{end}}$
$t = t + \Delta t$
**Energy**
$\partial T/\partial t + \nabla \cdot (\mathbf{u}T) - \nabla \cdot (\alpha \nabla T) = r$
under-relax, solve for $T$
**Momentum matrix**
$\mathbf{A} \cdot \mathbf{u} - \mathbf{b}$ Eq. (5.21)
under-relax matrix eqn
**Momentum predictor**
$\mathbf{A} \cdot \mathbf{u} - \mathbf{b} = -\nabla p$
***PIMPLE loop***
solve for $\mathbf{u}$
**Pressure equation**
$\nabla \cdot \frac{1}{A} \nabla p = \nabla \cdot \left[ \frac{\mathbf{H}(\mathbf{u})}{A} \right]$
solve for $p$
**Flux corrector**
$\phi_{\text{f}} := \mathbf{S}_{\text{f}} \cdot \left( \frac{\mathbf{H}(\mathbf{u})}{A} \right)_{\text{f}} - \left( \frac{|\mathbf{S}_{\text{f}}|}{A} \right)_{\text{f}} \nabla_{\mathbf{n}} p_{\text{f}}$
***PISO loop***
under-relax $p$
**Momentum corrector**
$\mathbf{u} := \frac{\mathbf{H}(\mathbf{u})}{A} - \frac{1}{A} \nabla p$
$p, \mathbf{u}, \phi_{\text{f}}, T$
**ENDWHILE**

## 5.22 Solving for energy

Examples with energy conservation in this chapter assume $c_V$ is constant to produce the transport equation for $T$ in Eq. (2.65). When $c_V$ cannot be assumed to be constant, it is common to solve an equation for internal energy $e$ instead, *e.g.* Eq. (2.60) with the material derivative replaced using Eq. (2.14)

$$\frac{\partial \rho e}{\partial t} + \nabla \bullet (\rho \mathbf{u} e) - \nabla \bullet \kappa \nabla T = 0. \tag{5.39}$$

This equation can be discretised to form a matrix equation $[\mathbf{A}][e] = [\mathbf{b}]$, from which $e$ can be solved. The challenge is that the diffusion term $\nabla \bullet \kappa \nabla T$ is not expressed in terms of $e$, so is discretised explicitly, which adversely affects convergence.

To improve convergence, an implicit term is introduced which is similar in form and scale to the problem term. For energy Eq. (5.39), we use $\nabla \bullet \alpha_e \nabla e \approx \nabla \bullet \kappa \nabla T$, where the diffusivity $\alpha_e = \kappa / c_V$ is calculated from the $c_V(T)$ function of $T$.

The extra term is both "added and subtracted" in implicit and explicit form. This has the effect of increasing the contributions to matrix coefficients, while cancelling out a large part of the explicit contribution from $\nabla \bullet \kappa \nabla T$, as illustrated below

$$\frac{\partial \rho e}{\partial t} + \nabla \bullet (\rho \mathbf{u} e) \underbrace{- \nabla \bullet \alpha_e \nabla e}_{\text{implicit}} + \underbrace{\nabla \bullet \alpha_e \nabla e - \nabla \bullet \kappa \nabla T}_{\text{explicit}} = 0.$$

The overall solution procedure involves first updating $e$ from $T$ from thermodynamic relationships, *e.g.* Eq. (2.62). The equation for $e$ is then solved, and the subsequent solution for $e$ is converted back to $T$.

If $c_V$ is expressed as a polynomial function of $T$, Eq. (2.64), then

$T$ is converted to $e$ using the analytical integral of Eq. (2.62)

$$e = \left[ \sum_{i=0}^{n} \frac{a_i}{i+1} T^{i+1} \right]_{T_{\mathrm{ref}}}^{T} + e_{\mathrm{ref}} . \tag{5.40}$$

## From temperature to energy

The conversion from $e$ to $T$ is more complex because $T$ cannot be made the subject of Eq. (5.40). Instead, it can be "inverted" using an iterative scheme such as the Newton-Raphson method,† which for this problem is:

$$T := T - \frac{e(T) - e}{\partial e / \partial T} = T - \frac{e(T) - e}{c_V(T)} . \tag{5.41}$$

$T$ is updated from the current $e$ using the evaluated $e(T)$ and $c_V(T)$ polynomials, Eq. (5.40) and Eq. (2.64), respectively. One iteration is often sufficient for the error $(e(T) - e)$ to fall within a prescribed tolerance, but further iterations of Eq. (5.41) can be applied if necessary.

## Boundary conditions

The boundary conditions for the energy equation are generally specified in terms of $T$. But since they must be applied to the variable being solved, they must be reposed in terms of $e$. A fixed value condition $T = T_{\mathrm{b}}$ is converted to an equivalent condition $e = e_{\mathrm{b}}$, *e.g.* by Eq. (5.40). A fixed gradient condition $\nabla_{\mathbf{n}} T_{\mathrm{b}}$ for $T$ is converted to a fixed gradient $\nabla_{\mathbf{n}} e_{\mathrm{b}}$ for $e$ by

$$\nabla_{\mathbf{n}} e_{\mathrm{b}} \approx c_V \nabla_{\mathbf{n}} T_{\mathrm{b}} . \tag{5.42}$$

† Isaac Newton, *De analysi per aequationes numero terminorum infinitas*, 1669 and Joseph Raphson, *Analysis aequationum universalis*, 1690.

## 5.23 Summary of algorithms and solvers

- The finite volume method produces sparse matrices, Sec. 5.1.
- Symmetric matrices have the same coefficients across the diagonal and advection generally produces an asymmetric matrix.
- The matrix equations are solved using iterative methods that reduce a residual to a specified tolerance, Sec. 5.4.
- Diagonal dominance is a sufficient condition for convergence, see Eq. (5.9) and Sec. 5.5.
- Under-relaxation may be required to maintain diagonal dominance, Sec. 5.6, in particular for steady-state solutions.

### Gauss-Seidel method

- Simple iterative method for solutions with cost $\propto N$, Sec. 5.2.
- Effective when the range of influence is short, *i.e.* $\boldsymbol{\Psi}$ in each cell is influenced by changes in close neighbouring cells.
- The symmetric variant and mesh renumbering may improve convergence, Sec. 5.8.

### Conjugate gradient method

- A "descent" method which solves a minimisation problem for cases with a longer range of influence, Sec. 5.14.
- Solved iteratively, which requires preconditioning to make it efficient, Sec. 5.16.
- Symmetric matrix: PCG with DIC preconditioning.
- Asymmetric matrix: PBiCGStab with DILU preconditioning.

## Geometric-algebraic multi-grid (GAMG)

- Well suited to equations with a long range of influence, in particular the pressure equation, Sec. 5.17.
- Simple algebraic agglomeration is corrected by a scaling based on minimisation for Laplacian-dominated equations, Sec. 5.18.

## Systems of equations

- Systems of equations are solved iteratively with solutions from one equation substituted into the next, Sec. 5.9.
- Mass and momentum conservation are coupled by a pressure equation and momentum and flux correctors, Sec. 5.10.

## Steady-state solution

- Steady flow solutions use the SIMPLE algorithm, Sec. 5.12.
- Equations are solved in an iterative sequence with under-relaxation required to ensure convergence, Sec. 5.13.

## Transient solution

- Transient solutions use a PISO loop, Sec. 5.19.
- Correction for non-orthogonality in the pressure equation may require a distinct corrector loop, Sec. 5.20.
- In the PISO loop, typical configurations solve: 2 pressure equations, each including 1 non-orthogonal corrector; or 3 pressure equations without any non-orthogonal correctors.
- The PIMPLE algorithm provides additional controls for transient and pseudo-transient simulations, Sec. 5.21.

# Chapter 6

# Introduction to Turbulence

*"But never have I been a blue calm sea, I have always been a storm."*

Fleetwood Mac, *Storms* (1979).

The previous chapters provide some general principles for CFD, from the governing equations and common models, to numerical methods, boundary conditions, algorithms and solvers.

Since the book is about general principles, it does not generally cover complex physical modelling. The exception is *turbulence* which need to be addressed since it is so common in fluid flow and therefore a critical part of most CFD simulations.

Turbulence is illustrated by a plume of smoke on the following page. The smoke rises vertically from the base, initially tracing a relatively straight, narrow path, characteristic of *laminar flow.* At some point, there is an abrupt transition to *turbulent flow* in which fluid particles follow curved paths which cross one another in a disorderly, irregular manner, causing $\mathbf{u}$ and $p$ to fluctuate randomly in time.

Turbulent flow evidently involves significant *mixing* of the fluid which is an important consideration in many CFD problems. For example, it directly affects the rate of pollutant dispersion and the rate of reaction of a chemical process. It causes heat to diffuse more rapidly. Similarly, mixing affects momentum diffusion and, consequently, the aerodynamic forces within a fluid and surrounding vehicles, wind turbines, buildings, *etc.*

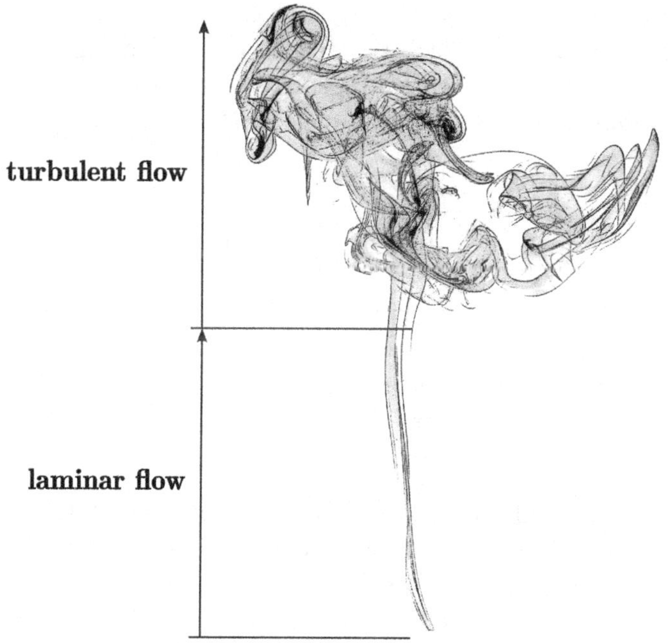


Turbulence is a vast subject with an abundance of textbooks for various levels of expertise. As an introduction to turbulence, this chapter aims to provide:

- a short description of the nature of turbulence;
- an understanding of its importance in analysis and particular challenges it presents in CFD;
- the basic concepts of turbulence modelling to lay the groundwork for the models presented in Chapter 7.

## 6.1 Reynolds experiment

Reynolds distinguished the laminar and turbulent flow regimes in his influential experiments in the early 1880s.† At that time, Poisueille's law predicted correctly that the resistance to flow in a pipe $\propto u$ for laminar flow at speed $u$, which typically occurs at small $u$ and/or diameter $D$. Reynolds wanted to understand why, at higher $u$ and/or $D$, resistance $\propto u^2$ (approximately).

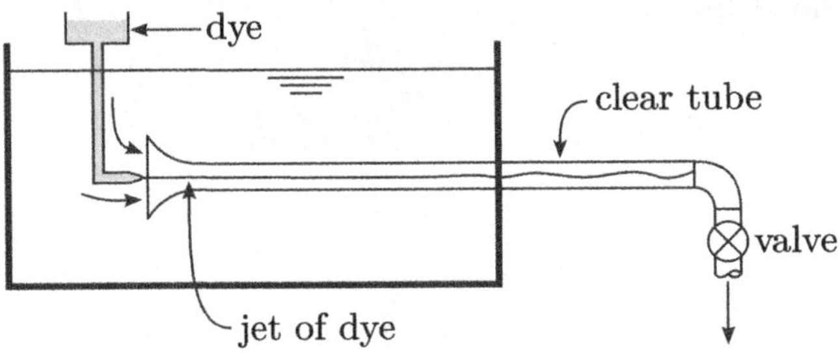


His experiment used a large, glass-walled tank filled with water. A glass tube with a flared intake passed through the tank and out through one wall. Water from the tank flowed along the tube at speed $u$, controlled by an outlet valve. A jet of liquid dye was injected at the inlet to the tube to visualise the flow.

The observed behaviour is shown in Reynolds's original drawings on the following page. At sufficiently low speed, the streak of dye followed a straight line along the tube, indicating laminar flow. The speed was increased in small steps until at some distance from the tube intake (typically, $30D$) the dye would mix with the water, rapidly filling the tube. This marked the *transition to turbulent flow.*

†Osborne Reynolds, *An experimental investigation of the circumstances which determine whether the motion of water shall be direct or sinuous, and of the law of resistance in parallel channels*, 1883.

Reynolds made the following important observations.

- Distinct *eddies* were visible during turbulent flow, see above.
- The transition speed was *very sensitive to disturbances* in the water entering the tube, even due to the temperature variations in the water.
- Before full transition was reached, intermittent "spots" of turbulence appeared and disappeared.

## Reynolds number

Reynolds argued that the transition is controlled by the Reynolds number, Re from Eq. (2.68), using the characteristic length $L = D$ in a pipe. His deduction was based on the idea of scale similarity, introduced in Sec. 2.21.

He observed that with minimal disturbance in the tank, the transition to turbulent flow occurred at $\mathrm{Re} = 13000$. When disturbances were evidently present, the transition occurred at $\mathrm{Re} \approx 2000$. An updated view for pipe flow is that for $\mathrm{Re} \lesssim 2000$, all disturbances will decay preventing the onset of turbulent flow. For $\mathrm{Re} \gtrsim 2000$, transition depends on initial disturbances and the roughness of the pipe wall.

## 6.2 A picture of turbulence

In turbulent flow, the fluid follows irregular curved paths, known as eddies. Eddies are generally illustrated by curved arrows, which does little to reveal the true nature of turbulence.

Instead, we can build a picture of turbulence from: photographs and videos from experiments, examples from engineering and everyday experiences; and, from images and animations from detailed computer simulations.

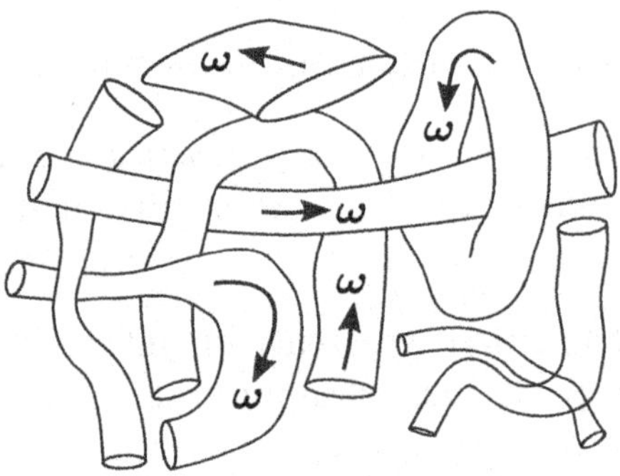


Generally, we see a mass of intertwined, flow structures of many different sizes. The structures move and rotate continuously. Through interactions with one another and the main flow field, they change shape and size rapidly.

The structures indicate regions of *vorticity* $\boldsymbol{\omega}$ in the fluid, as defined in Sec. 2.11. It is vorticity, not velocity, which is often used to describe turbulence. For example, Davidson† describes *turbulence* as "a spatially complex distribution of vorticity which advects itself in a chaotic manner", and a *turbulent eddy* as "a blob, tube or sheet of vorticity and the associated flow."

†Peter Davidson, *Turbulence: an introduction for scientists and engineers*, 2004.

## Vorticity generation

Vorticity is primarily generated at solid boundaries and can be categorised as "fast" or "slow".† Fast generation is localised to regions of relative acceleration between the solid and fluid, *e.g.* at the leading edge in flow over a semi-infinite flat plate or at the lip at the exhaust of the air cannon in Sec. 2.11.

Slow generation occurs over longer sections of boundary surface, *e.g.* the wall of a pipe with internal fluid flow. It is caused by the tangential pressure gradient at the surface.

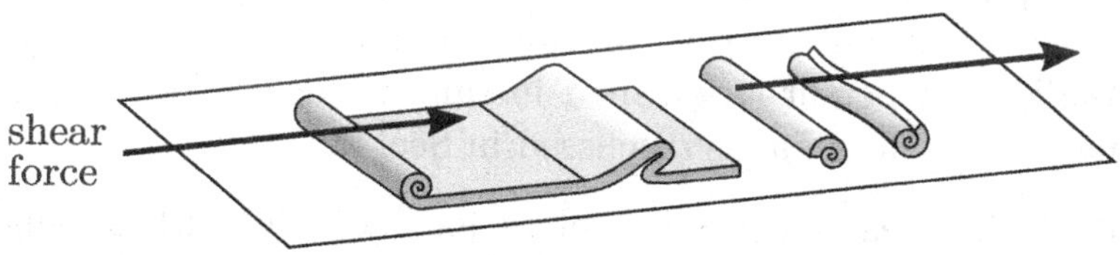


Since vorticity is mostly generated at solid boundaries, we can try to imagine the how vortex structures might evolve. Consider a layer of a viscoelastic fluid on a flat surface — something like a sticky adhesive or putty. When shear is applied, the material tends to rotate rather than slide horizonatally.

Sections of material can break off and form tubes which roll across the surface. They tend to extend along their length and bend under uneven shear.

When we apply shear by hand, we can feel that the rotation is initiated within the uneven points, *e.g.* around the finger joints. In a similar way, the onset of turbulence is sensitive to surface roughness with the transition from laminar to turbulent flow occurring at higher Re for smoother surfaces.

---

†Bruce Morton, *The generation and decay of vorticity*, 1984.

## 6.3 Vorticity transport

For an incompressible fluid (with other simplifying assumptions), vorticity obeys the transport equation on the following page, Eq. (6.2). This is a typical advection-diffusion equation, similar to Eq. (2.65) for heat, which is expressed in terms of the local time derivative and advection in conservative form by

$$\frac{\partial \boldsymbol{\omega}}{\partial t} + \underbrace{\nabla \bullet (\mathbf{u}\,\boldsymbol{\omega})}_{\text{advection}} - \underbrace{\nu \nabla^2 \boldsymbol{\omega}}_{\text{diffusion}} = \underbrace{\boldsymbol{\omega} \bullet (\nabla \mathbf{u})}_{\text{source}}. \tag{6.1}$$

Notably, Eq. (6.1) does not include a term in $p$. This is in contrast with conservation of (linear) momentum which can redistribute a perturbation in $\mathbf{u}$ instantaneously across all the domain through $-\nabla p$, as discussed in Sec. 2.22.

Instead, like heat, vorticity evolves locally *only*, with a range of influence limited by advective and diffusive transport, as discussed on page 50.

Advection of vorticity is clearly illustrated by the smoke ring shown in Sec. 2.11. Diffusion occurs by viscous torques transferring angular momentum between fluid elements.

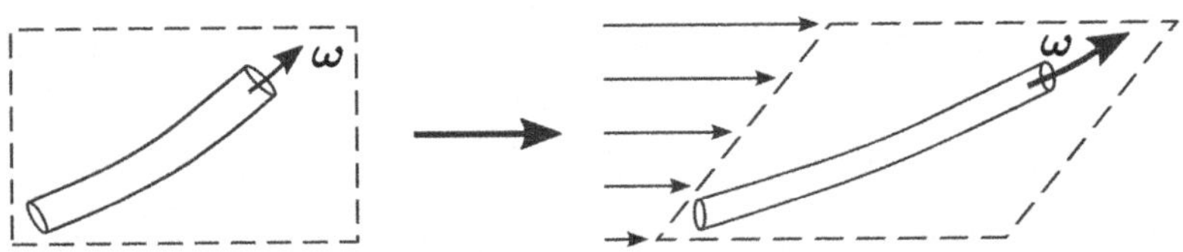


The source of vorticity, $\boldsymbol{\omega} \bullet (\nabla \mathbf{u})$ is due to vortices changing shape under the influence of a velocity gradient $\nabla \mathbf{u}$. If a vortex is stretched, *e.g.* under shear as shown above, its radius decreases, so angular velocity, and thus $\boldsymbol{\omega}$, increases. Similarly, $\boldsymbol{\omega}$ decreases if the vortex is compressed.

## The vorticity transport equation

For incompressible flow with constant $\nu$ (and zero, or constant, body force), vorticity obeys the following transport equation:

$$\boxed{\frac{D\boldsymbol{\omega}}{Dt} - \boldsymbol{\omega} \bullet (\nabla \mathbf{u}) - \nu \nabla^2 \boldsymbol{\omega} = \mathbf{0}} \tag{6.2}$$

The equation is derived from momentum conservation of a homogeneous, incompressible, Newtonian fluid with $\nu =$ constant and ignoring body forces, Eq. (2.49). Noting that $\nabla \bullet (\mathbf{uu}) \equiv \mathbf{u} \bullet \nabla \mathbf{u}$ due to Eq. (2.46) leads to

$$\frac{\partial \mathbf{u}}{\partial t} + \mathbf{u} \bullet (\nabla \mathbf{u}) + \nu \nabla^2 \mathbf{u} = -\nabla p. \tag{6.3}$$

The vorticity equation is *derived from the curl ($\nabla \times$) of Eq. (6.3)* combined with the vorticity definition Eq. (2.37). The first term is $\nabla \times (\partial \mathbf{u}/\partial t) \equiv \partial \boldsymbol{\omega}/\partial t$.

Replacing $\mathbf{a}$ and $\mathbf{b}$ by $\mathbf{u}$ in Eq. (2.72d) and applying Eq. (2.37) gives

$$\mathbf{u} \bullet \nabla \mathbf{u} \equiv \frac{1}{2} \nabla (\mathbf{u} \bullet \mathbf{u}) - \mathbf{u} \times \boldsymbol{\omega}. \tag{6.4}$$

The curl of the second term in Eq. (6.3) is expressed by the curl of Eq. (6.4) with Eq. (2.75a), Eq. (2.73a), Eq. (2.75e) and Eq. (2.46), leading to

$$\nabla \times (\mathbf{u} \bullet \nabla \mathbf{u}) \equiv \mathbf{u} \bullet (\nabla \boldsymbol{\omega}) - \boldsymbol{\omega} \bullet (\nabla \mathbf{u}). \tag{6.5}$$

The curl of the third term $\nabla \times (\nu \nabla^2 \mathbf{u}) \equiv \nu \nabla^2 \boldsymbol{\omega}$ by Eq. (2.74f). The curl of the fourth term $\nabla \times \nabla p = 0$ by Eq. (2.75a).

Combining all the terms and applying the material derivative Eq. (2.14), $D\boldsymbol{\omega}/Dt \equiv \partial \boldsymbol{\omega}/\partial t + \mathbf{u} \bullet (\nabla \boldsymbol{\omega})$ to Eq. (6.5), leads to Eq. (6.2).

## 6.4 Boundary layers

Boundary layers† are regions of fluid formed along solid boundaries in which the velocity **u** varies: from zero at the boundary (no-slip condition, Sec. 4.4); to a value largely unaffected by the proximity of the boundary, determined by the flow conditions.

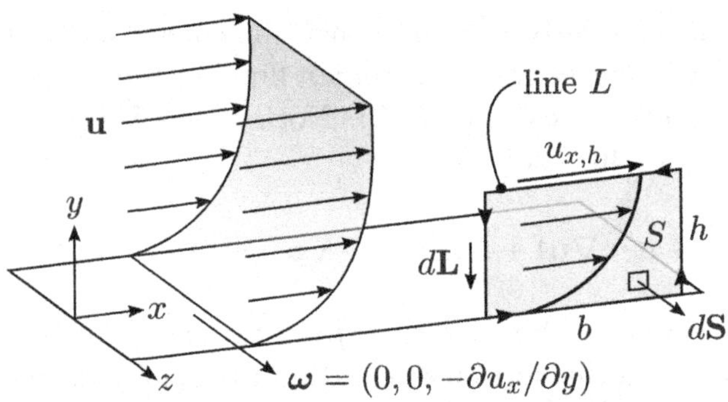


The figure above shows a boundary layer for flow in the $x$-direction at speed $u_x$, along a flat solid boundary oriented in the $y$-normal direction. At the boundary surface, the vorticity $\omega_z = -\partial u_x/\partial y$ is significant.

Vorticity can be shown over a planar section of the boundary layer of width $b$ and height $h$ (above, right). Applying Stokes's theorem, Eq. (2.39), the integral $= -u_{x,h}\, b$ along the upper line (with zero along the wall and the verticals sides). The average vorticity over plane area $b \times h$ is therefore $\omega_z = -u_{x,h}/h$.

Boundary layers are the main source of vorticity for turbulence. Turbulence occurs when instabilities, *e.g.* induced by roughness of the boundary surface, cause the vorticity to become chaotic, sustained by a sufficiently high Re.

---

†Ludwig Prandtl, *Über Flüssigkeitsbewegung bei sehr kleiner Reibung*, 1904.

The growth of boundary layers is related to vorticity transport. For flow over a flat plate, vorticity generated at the leading edge is advected by the flow, while diffusing away from the plate.

$$\frac{\partial \boldsymbol{\omega}}{\partial t} + \nabla \bullet (\mathbf{u}\,\boldsymbol{\omega}) - \nu \nabla^2 \boldsymbol{\omega} = \boldsymbol{\omega} \bullet (\nabla \mathbf{u}).$$

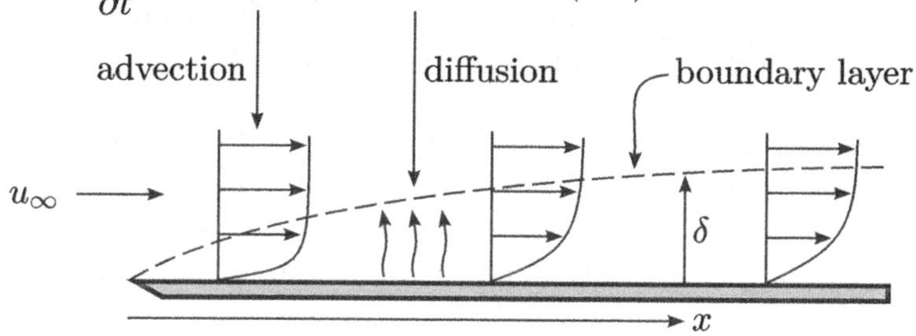


Vorticity propagates by diffusion by a distance $\delta \sim \sqrt{\nu t}$ in time $t$, see Sec. 2.22. In that time, it is advected a distance $x = u_\infty t$, where $u_\infty$ is the freestream flow speed. Comparing the distances over the same $t$, the boundary layer thickness is

$$\frac{\delta}{x} \propto \sqrt{\frac{\nu}{u_\infty x}} = \sqrt{\frac{1}{\text{Re}}}. \tag{6.6}$$

The relation $\delta/x = A\,\text{Re}^{-1/2}$ is suitable for *laminar* boundary layers, with coefficient $A$ depending on the definition of $\delta$. Data and analysis, including *e.g.* the Blasius solution[†], indicate $A \approx 5$ in the case of the "99% thickness", *i.e.* the distance from the wall where velocity reaches 99% of its asymptotic value.

In *turbulent* boundary layers, the diffusion front advances more rapidly due to mixing, see Sec. 6.11. As a result, $\delta/x$ is relatively insensitive to Re, *e.g.* the analytical solution $\delta/x \approx 0.37\,\text{Re}^{-1/5}$, based on a one-seventh (1/7th) power law for the velocity profile.

[†]Paul Richard Heinrich Blasius, *Grenzschichten in Flüssigkeiten mit kleiner Reibung*, 1908.

## 6.5 Boundary layer separation

Boundary layers provide the main source of vorticity for turbulence as discussed in Sec. 6.4. A boundary layer breaks away from the boundary when it reaches the end of the surface or by *separation* before reaching the end. Vorticity and turbulence are thereby swept into regions of fluid away from solid boundaries.

Flow around a cylinder illustrates boundary layer separation, vorticity and turbulence. The fluid flows at uniform velocity $u_x = u_\infty$ upstream of the cylinder. It decelerates to *stagnation*, $u_x = 0$, at point A on the surface (in the $-x$-normal direction).

High pressure at A introduces a *favourable* pressure gradient $\partial p/\partial\theta < 0$ which increases the flow speed $u_\theta$ around the cylinder towards B, developing a boundary layer in the process.

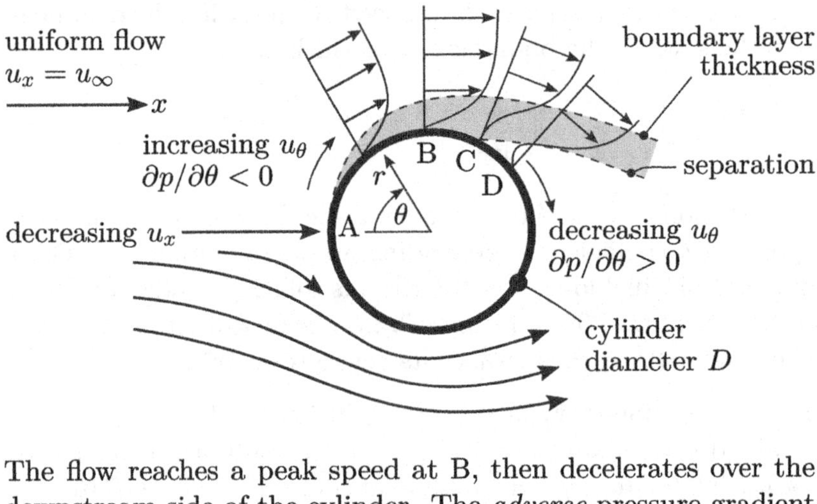


The flow reaches a peak speed at B, then decelerates over the downstream side of the cylinder. The *adverse* pressure gradient $\partial p/\partial\theta > 0$ causes $u_\theta$ to decrease. At some point C, the velocity gradient can reach $\partial u_\theta/\partial r = 0$. Beyond C, the boundary layer can *separate* such that $u_\theta < 0$ along its profile, see point D.

Boundary layer separation in a cylinder depends on Reynolds number Re, Eq. (2.68), using $L = D$ and $U = u_\infty$. For Re $<$ 5, there is no separation, with the flow exhibiting a pattern downstream that mirrors the upstream flow.†

For $5 < \text{Re} < 40$, the boundary layer separates with its vorticity sustaining a pair of vortices attached to the rear of the cylinder.

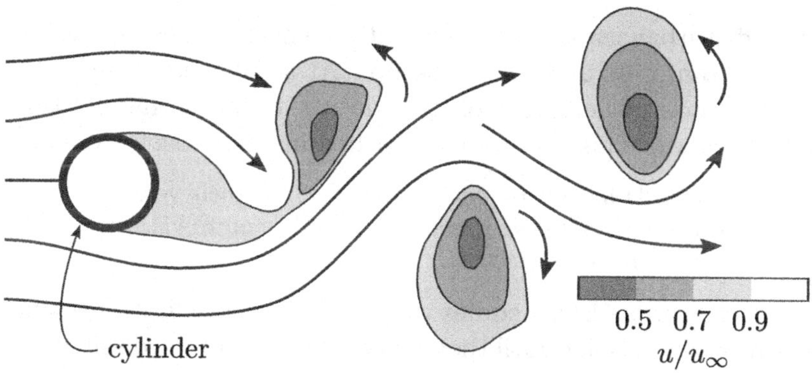


At Re $\approx 100$, vorticity is released downstream as vortices break off from the cylinder in a periodic manner known as the *Kármán vortex street*, shown above.

At Re $\approx 400$, the vorticity starts to become become chaotic, with turbulence beginning to appear in the vortices. At Re $\approx 10^5$, the entire wake region becomes turbulent.

The frequency of vortex shedding is characterised by another dimensionless number from Eq. (2.68), the Strouhal number Sr. For $200 < \text{Re} < 10^5$, experiments show‡ Sr $\approx 0.2$, where $\mathcal{T}$ is the period at which the vortex pattern repeats.

†Sadatoshi Taneda, *Experimental investigation of the wakes behind cylinders and plates at low Reynolds numbers*, 1956.

‡Vincenc Strouhal, *Über eine besondere Art der Tonerregung*, 1878 and Anatol Roshko, *On the development of turbulent wakes from vortex streets*, NACA Report 1191, 1954.

## 6.6 Scales of turbulence

In Sec. 6.2, turbulence was pictured as a mass of intertwined eddies which are rotating and distorting. The rotation, *i.e.* vorticity, mostly originates at solid boundaries and propagates locally (Sec. 6.3) through boundary layers (Sec. 6.4), which can separate and shed into the flow field (Sec. 6.5).

Flow disturbances trigger instabilities which cause vortices to stretch, compress and/or 'break up'. At a sufficiently high Re, coherent flow structures rapidly disintegrate into a mass of turbulent eddies. This process is described in Richardson's poem:[†]

> Big whorls have little whorls that feed off their velocity,
> and little whorls have lesser whorls and so on to viscosity
> — in the molecular sense.

In other words, large eddies created by instabilities become progressively smaller until they reach a size where the dissipation of their kinetic energy due to (molecular) viscosity is significant. The loss of kinetic energy causes these eddies to "die out" before they can become any smaller.

### Kolmogorov microscales

The *Kolmogorov microscales*[‡] describe the smallest scales that can exist in turbulent flow. They can be derived from heuristic arguments as follows. Energy dissipates as heat at a rate of $2\nu(\operatorname{dev}\mathbf{D}) \mathbin{:} \nabla\mathbf{u}$ per unit mass, obtained from Eq. (2.58) and switching from 'per unit volume' by replacing $\mu$ by $\nu$.

Using scale similarity arguments (Sec. 2.21) the average rate of

---

[†]Lewis Fry Richardson, *Weather prediction by numerical process*, 1922.

[‡]Andrey Nikolaevich Kolmogorov, *The local structure of turbulence in incompressible viscous fluid for very large Reynolds numbers*, first published in Russian in *Dokl. Akad. Nauk SSSR* **30**, 1941.

dissipation of energy $\varepsilon$ can then be estimated for the smallest scales, characterised by length $\eta$ and speed $v$, to be

$$\varepsilon \sim \nu(v/\eta)^2, \tag{6.7}$$

where "$\sim$" means "of the order of magnitude".

Dissipation occurs when the Reynolds number, based on the Kolmogorov scales, is of the order unity, *i.e.* $v\eta/\nu \sim 1$. Combined with Eq. (6.7) this yields the Kolmogorov microscales below.

$$\text{length} \qquad \eta \sim \left(\frac{\nu^3}{\varepsilon}\right)^{1/4} \tag{6.8a}$$

$$\text{velocity} \qquad v \sim (\nu\varepsilon)^{1/4} \tag{6.8b}$$

$$\text{time} \qquad \eta/v \sim \left(\frac{\nu}{\varepsilon}\right)^{1/2} \tag{6.8c}$$

At the largest scales, eddies generated from the mean flow can be characterised by a length $l$ and flow speed $u$ (of the scale of turbulent fluctuations). The kinetic energy per unit mass of these eddies $\sim u^2$.

Experiments show the lifetime of an eddy is of the order of the time for one revolution, the *turn-over* time $\sim l/u$. Therefore the rate of transfer of kinetic energy from larger to smaller eddies is $\sim u^2/(l/u) = u^3/l$.

This rate of transfer of energy between scales must match the rate of dissipation, Eq. (6.7). Equating the two energy rates yields the correspondence between the Kolmogorov scales and the largest scales in terms of a large scale $\mathrm{Re} = ul/\nu$:

$$\eta \sim l\,\mathrm{Re}^{-3/4}; \quad v \sim u\,\mathrm{Re}^{-1/4}; \quad \eta/v \sim (l/u)\,\mathrm{Re}^{-1/2}. \tag{6.9}$$

## 6.7 Energy cascade

The process of large eddies becoming smaller is introduced in Sec. 6.6. The process involves a transfer of kinetic energy from larger to smaller eddies, known as a the *energy cascade.*

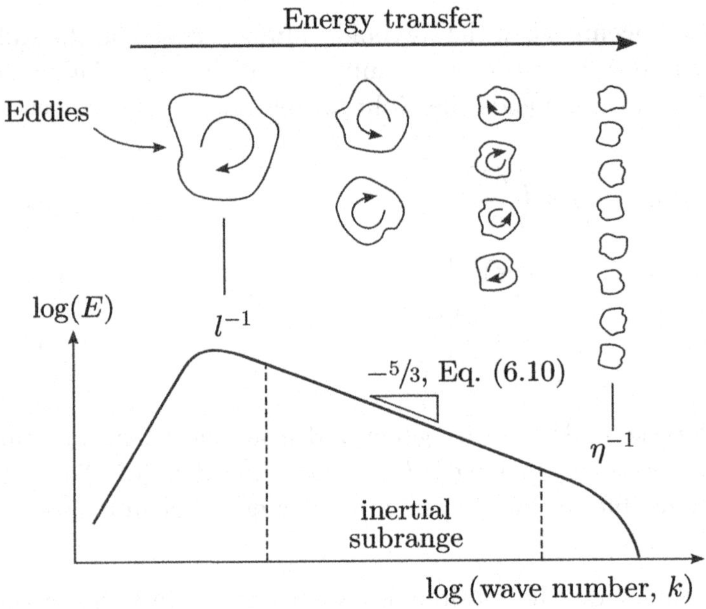


The variation in kinetic energy with eddy size is often illustrated by the energy spectrum graph above. The horizontal axis is wave number $k$,[†] which represents the number of eddies per unit length, *i.e.* $\propto r^{-1}$ where $r$ = eddy size.

The cascade starts with the largest eddies of size $l$ with the highest level of turbulent kinetic energy per unit mass (TKE). In the range, $k \to k + dk$, TKE $= E(k)\, dk$.

[†]The symbol $k$, is commonly used in science to denote wave number; it should not be confused with its later use to denote turbulent kinetic energy.

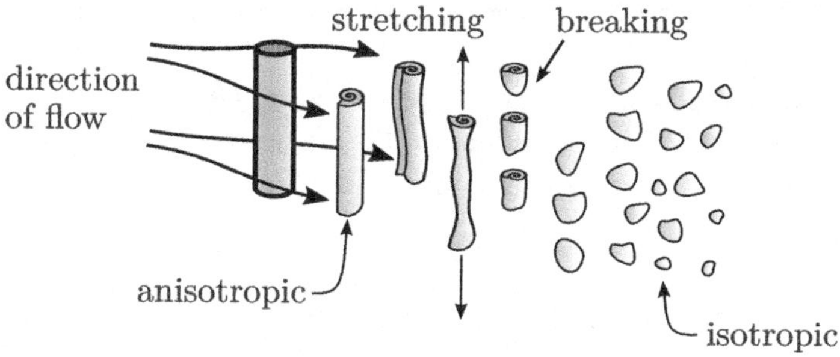


Large eddies are usually *anisotropic* due to the way turbulence is generated. For example, flow past a cylinder causes shedding of vortices whose shape are similar to the cylinder, *i.e.* longer in the direction of the cylinder axis.

Once moving through open space, eddies quickly stretch, bend, rotate and break, so that quite soon they become "blobs" of vorticity. The turn-over time, initially associated with shedding, decreases with eddy-size — by a factor of $\mathrm{Re}^{1/2}$ at the smallest scales, according to Eq. (6.9). Any dominant frequencies, *e.g.* those due to shedding, are thereby quickly lost.

Kolmogorov's hypothesis was that the turbulence then becomes *isotropic* and that $E(k)$ is then only a function of $\varepsilon$ and $r$ when Re is large. From this hypothesis emerges his *five-thirds law* for the *inertial subrange* $\eta \ll r \ll l$

$$E(k) = \alpha\, \varepsilon^{2/3} k^{-5/3}, \tag{6.10}$$

where $\alpha \approx 1.5$ for $\mathrm{Re} \to \infty$. The exponents of $^2/_3$ and $^5/_3$ are chosen to match the dimensions of $E$ which are $L^3/\mathcal{T}^2$, and noting $r \propto k^{-1}$.

Experimental data generally supports Eq. (6.10) — remarkably, given its simplicity.

## 6.8 The cost of simulating turbulence

The Kolmogorov microscales are the smallest scales of length, velocity and time in turbulence. The fact that the scales are so small has important consequences for CFD.

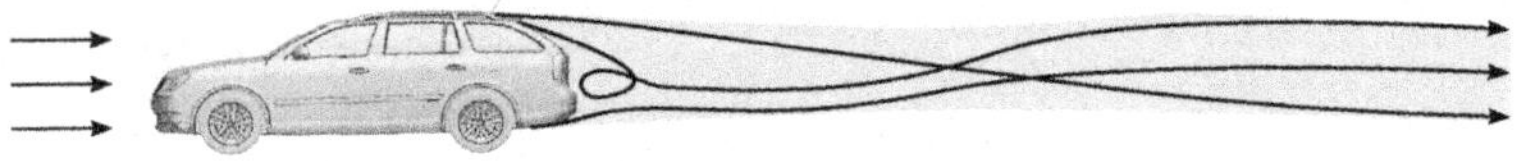


In automotive external aerodynamics, boundary layers develop from the front of the vehicle which separate towards the rear. The flow recirculates, forming a wake region with high levels of turbulence.

The Kolmogorov length scale $\eta \sim 0.1\,\mathrm{mm}$, with a time scale of $\sim 1\,\mathrm{ms}$, based on an estimated turbulent energy dissipation rate $\varepsilon \sim 10\,\mathrm{m^2s^{-3}}$ and $\nu = 1.5 \times 10^{-5}\,\mathrm{m^2s^{-1}}$ for air.

To resolve the smallest eddies with CFD, a mesh with typical cell length $\sim 0.01$ mm would then be required. Since we would need to capture a wake region of volume $\sim 10$ m$^3$, the corresponding mesh would consist of $\sim 10/(10^{-5})^3 = 10^{16}$ cells.

In automotive external aerodynamics today (2022), some of the largest simulations run with meshes of $\sim 10^8$ cells on $\sim 1000$ processor cores. These mesh sizes are clearly several orders of magnitude smaller than those required for *direct numerical simulation* (DNS) of turbulence at the smallest scales.

An important consequence of using a mesh that does not resolve the smallest scales is that the main mechanism of dissipating turbulent energy is not captured. This affects the rate of energy transfer within the energy cascade.

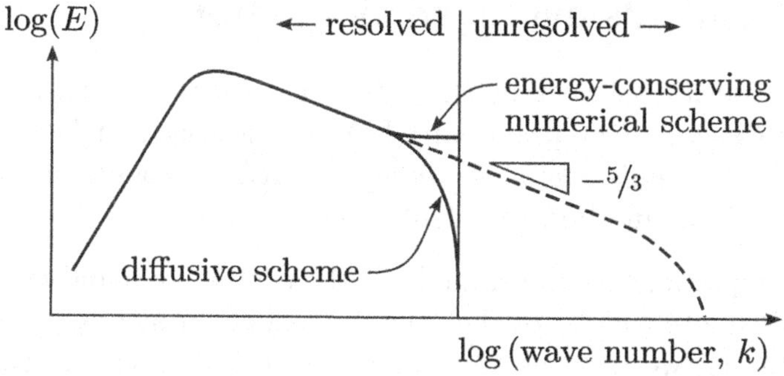


If the energy dissipation is neither captured nor introduced by diffusive numerical schemes, excess energy inevitably "backs up", causing over-prediction of the turbulence in the resolved eddies. Similarly, turbulence is under-predicted with diffusive schemes that introduce excessive numerical dissipation.

*Large-eddy simulation* (LES) instead uses modelling for the unresolved scales while capturing the large eddies using accurate numerical schemes. Effective models dissipate the energy in the unresolved scales at a realistic rate, *e.g.* by the five-thirds law.

Turbulent fluctuations are predicted within the resolved range of wave numbers, *i.e.* corresponding to lower frequencies. LES is therefore useful for problems involving excitation at those frequencies, *e.g.* flow-induced vibration, jet breakup and noise.

The simulations are still relatively costly, however, since they are inherently transient, requiring strong convergence within each time step for accuracy. Mesh sizes are still relatively large and long simulation times are needed to generate reliable time-averaged properties.

We therefore need a faster method to calculate turbulence, suitable for flows which are steady (or reasonably steady).

## 6.9 Reynolds-averaged simulation

The computational cost of DNS and, to a lesser extent, LES is too great for most practical CFD, as discussed in Sec. 6.8. Instead, a *Reynolds-averaged simulation* (RAS) provides a much more affordable method to calculate turbulence.

It solves equations for "averaged" field variables to avoid resolving small fluctuations. Rather than consider an average over a time interval, we imagine the same flow repeated several times under nominally the same initial conditions (2 examples below).

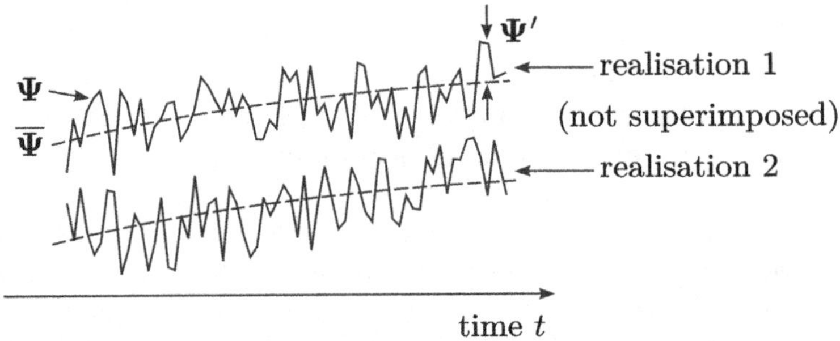


Solutions vary due of differences in initial conditions and the chaotic nature of turbulence. The *ensemble* average calculates the mean solution $\overline{\boldsymbol{\Psi}}$ for multiple *realisations* of the same flow.

Each field, *e.g.* $\boldsymbol{\Psi}$, is decomposed into the averaged field $\overline{\boldsymbol{\Psi}}$ and field of random fluctuations $\boldsymbol{\Psi}'$, according to

$$\boldsymbol{\Psi}(t) = \overline{\boldsymbol{\Psi}} + \boldsymbol{\Psi}'(t). \tag{6.11}$$

We can apply this decomposition to fields in the momentum conservation Eq. (2.27) with stress $\boldsymbol{\sigma}$ split into $p$ and $\boldsymbol{\tau}$ as follows:

$$\frac{\partial \rho \mathbf{u}}{\partial t} + \nabla \cdot (\rho \mathbf{u}\mathbf{u}) = \nabla \cdot \boldsymbol{\tau} - \nabla p + \rho \mathbf{b}. \tag{6.12}$$

Let us assume constant $\rho$ and that the body force $\mathbf{b}$ is not subject to turbulent fluctuations. Splitting the remaining fields, *i.e.* $\mathbf{u}$, $\boldsymbol{\tau}$

and $p$, into instantaneous and fluctuating components according to Eq. (6.11), and taking the ensemble average of Eq. (6.12), yields the following:

$$\frac{\partial \rho \overline{\mathbf{u}}}{\partial t} + \nabla \boldsymbol{\cdot} (\rho \overline{\mathbf{u}}\overline{\mathbf{u}}) + \nabla \boldsymbol{\cdot} (\rho \overline{\mathbf{u}'\mathbf{u}'}) = \nabla \boldsymbol{\cdot} \overline{\boldsymbol{\tau}} - \nabla \overline{p} + \rho \mathbf{b} \tag{6.13}$$

The averaged Eq. (6.13) is derived using the following relations for general fields $\boldsymbol{\Psi}$ and $\mathbf{a}$: $\overline{\boldsymbol{\Psi}'} = 0$; $\overline{\overline{\boldsymbol{\Psi}}} = \overline{\boldsymbol{\Psi}}$; $\overline{\overline{\boldsymbol{\Psi}}\mathbf{a}} = \overline{\boldsymbol{\Psi}}\overline{\mathbf{a}}$; $\overline{\boldsymbol{\Psi} + \mathbf{a}} = \overline{\boldsymbol{\Psi}} + \overline{\mathbf{a}}$. Relations for averaged derivatives are: $\overline{\partial \boldsymbol{\Psi}/\partial t} = \partial \overline{\boldsymbol{\Psi}}/\partial t$; $\overline{\nabla \boldsymbol{\Psi}} = \nabla \overline{\boldsymbol{\Psi}}$; $\overline{\nabla \boldsymbol{\cdot} \boldsymbol{\Psi}} = \nabla \boldsymbol{\cdot} \overline{\boldsymbol{\Psi}}$.

## Reynolds stress

The terms for mean quantities in Eq. (6.13) are the same as in Eq. (6.12). The difference is that Eq. (6.13) includes the additional $\nabla \boldsymbol{\cdot} (\rho \overline{\mathbf{u}'\mathbf{u}'})$ term containing fluctuations $\mathbf{u}'$.

The argument of this divergence derivative is a tensor known as the *Reynolds stress*[†]

$$\boxed{\boldsymbol{\sigma}^{\mathrm{t}} = -\rho \overline{\mathbf{u}'\mathbf{u}'}} \tag{6.14}$$

Substituting Eq. (6.14) in Eq. (6.13) eliminates fluctuation terms. The remaining equation is in terms of averaged properties only, so we can dispose of the average notation $(\bar{\ })$ to give

$$\frac{\partial \rho \mathbf{u}}{\partial t} + \nabla \boldsymbol{\cdot} \rho \mathbf{u}\mathbf{u} = \nabla \boldsymbol{\cdot} \left(\boldsymbol{\tau} + \boldsymbol{\sigma}^{\mathrm{t}}\right) - \nabla p + \rho \mathbf{b}. \tag{6.15}$$

This ensemble-averaged equation is the same form as Eq. (6.13) but with the addition of $\boldsymbol{\sigma}^{\mathrm{t}}$. Solving this Reynolds-averaged equation is the key to low cost CFD with turbulence — but it requires a *model* for the additional unknown $\boldsymbol{\sigma}^{\mathrm{t}}$.

[†]Osborne Reynolds, *On the dynamical theory of incompressible viscous fluids and the determination of the criterion*, 1895.

## 6.10 The nature of viscosity

Before discussing turbulence further, it is useful to examine the origins of viscosity. Viscosity is introduced in Sec. 2.12 of this book as part of the Newtonian constitutive model. The model was originally phenomenological, but was later derived directly from kinetic theory which describes a gas as a number of submicroscopic *particles*, *e.g.* atoms or molecules, in random motion.

The kinetic view of viscosity imagines a fluid in two dimensions, $x$ and $y$, subjected to a shear force in the $x$-direction. Although the mean flow is in the $x$-direction, particles move in the $y$-direction due to random fluctuations with a *mean speed* $\overline{v}$.

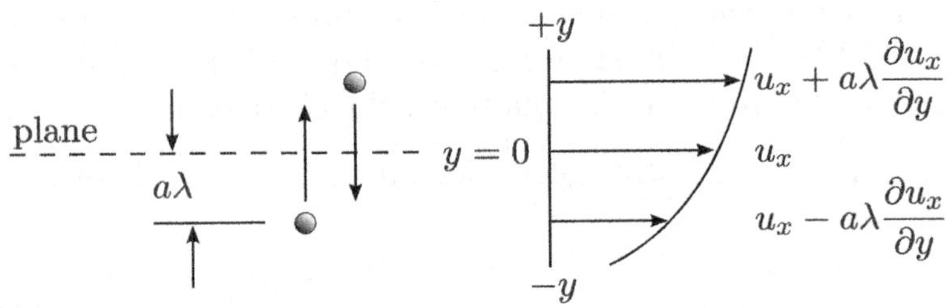


Consider a plane at $y = 0$. A particle will pass through the plane if its path towards it is not interrupted by a collision which sends it moving away from the plane. Particles passing through the plane arrive from an average distance $a\lambda$, where $a$ is some factor of the average distance travelled by a moving particle between successive collisions, the *mean free path* $\lambda$.

From kinetic theory, the mass flow rate of particles passing through a surface of unit area $= \rho\overline{v}/4$. The mean $x$-velocity of particles crossing the plane from the $+y$-direction is $u_x + a\lambda\partial u_x/\partial y$; similarly from the $-y$-direction, it is $u_x - a\lambda\partial u_x/\partial y$.

The net $x$-momentum of the particles, positive on the $+y$ side of the plane is then

$$\frac{1}{4}\rho\bar{v}\left(u_x + a\lambda\frac{\partial u_x}{\partial y}\right) - \frac{1}{4}\rho\bar{v}\left(u_x - a\lambda\frac{\partial u_x}{\partial y}\right) = \frac{1}{2}\rho\bar{v}a\lambda\frac{\partial u_x}{\partial y}. \quad (6.16)$$

The net momentum is equivalent to the shear stress on the $+y$ side of the plane, $\tau_{yx} = \mu\,\partial u_x/\partial y$, as described in Sec. 2.6. By comparison with Eq. (6.16), the dynamic viscosity $\mu = \rho\bar{v}a\lambda/2$ in terms of molecular properties. The kinematic viscosity is

$$\nu = \frac{a}{2}\bar{v}\lambda. \quad (6.17)$$

The original analysis of Maxwell† uses $a = 2/3$ in Eq. (6.17). It was later recognised the average distance described by $a$ was larger due to *persistence of velocities*, *i.e.* a particle will sometimes maintain a path towards the plane after a collision.

A more thorough analysis‡ begins with the Boltzmann equation and applies the Chapman–Enskog expansion to first order in Knudsen number

$$\mathrm{Kn} = \frac{\lambda}{L}. \quad (6.18)$$

The analysis derives the conservation laws with Newtonian and Fourier constitutive models, *i.e.* Eq. (2.19), Eq. (2.51), Eq. (2.41) and Eq. (2.54). Treating particles as rigid spheres and collisions as elastic, yields a value $a = 1.016 \times 5\pi/16 \simeq 1$, leaving a simple expression for viscosity which is

$$\boxed{\nu \simeq \frac{1}{2}\bar{v}\lambda} \quad (6.19)$$

---

†James Clerk Maxwell, *Illustrations of the dynamical theory of gases. Part I. On the motions and collisions of perfectly elastic spheres*, 1860.

‡see Sydney Chapman and Thomas Cowling, *The mathematical theory of non-uniform gases (3rd ed.)*, 1970.

## 6.11 Turbulent mixing

Turbulent flow is characterised by significant mixing of fluid eddies as the Reynolds experiment in Sec. 6.1 shows. CFD simulations generally need to accommodate turbulent mixing since it influences the diffusion of mass, momentum and energy. While the fluid mixing by mass diffusion itself can be important, the effect on momentum diffusion is often critical because it impacts the calculation of viscous forces and, thus, the flow itself.

### Eddy viscosity

Boussinesq was the first to devise a model for turbulence. He recognised the similarity between the random motion of both eddies in a turbulent fluid and particles at a molecular scale.

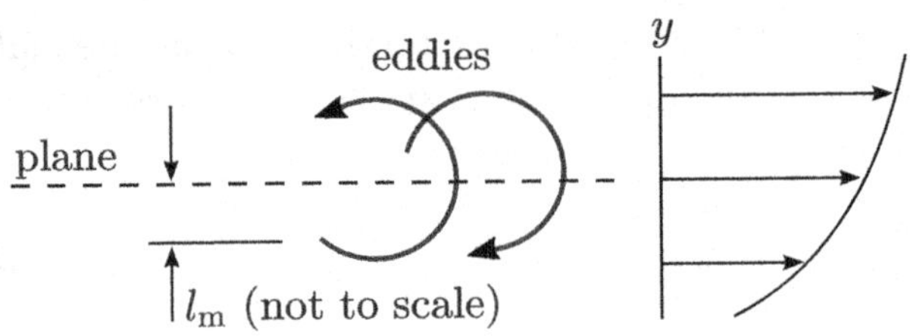


By analogy to kinetic theory in Sec. 6.10, shear stresses due to turbulence are caused by the net momentum, tangential to a plane, due to the motion of eddies. Boussinesq related this shear stress to the velocity gradient through an *eddy viscosity* $\nu_{\mathrm{t}}$.†

He presented the turbulent stress $\boldsymbol{\sigma}^{\mathrm{t}}$ in tensor form, including a pressure. Kinetic theory relates pressure to fluctuations‡ in particle velocity $\mathbf{v}'$ by $p = \rho\overline{\mathbf{v}'^2}/3$; the kinetic energy associated

† Joseph Boussinesq, *Essai sur la théorie des eaux courantes*, 1877.

‡ Daniel Bernoulli, *Hydrodynamica, sive de viribus et motibus fluidorum commentarii*, 1738.

with the fluctuations is $\rho\overline{\mathbf{v}'^2}/2$. Applying the same argument to velocity fluctuations $\mathbf{u}'$ due to turbulence, leads to a turbulent "pressure" $= {}^2\!/\!_3\rho k$, where $k = \overline{\mathbf{u}'^2}/2$ is the *turbulent kinetic energy* per unit mass.

By analogy with the Newtonian fluid model Eq. (2.41), the eddy viscosity model of Boussinesq, incorporating $k$, is

$$\boxed{\boldsymbol{\sigma}^{\mathrm{t}} = 2\rho\nu_{\mathrm{t}}\,\mathrm{dev}\,\mathbf{D}; \quad \boldsymbol{\sigma}^{\mathrm{t}} = \boldsymbol{\tau}^{\mathrm{t}} - \frac{2}{3}\rho k\,\mathbf{I}} \tag{6.20}$$

where $\boldsymbol{\tau}^{\mathrm{t}}$ is the viscous component of Reynolds stress and $\mathbf{D}$ is the deformation rate tensor defined in Eq. (2.33). Inevitably Eq. (6.20) and Eq. (2.41) closely resemble one another.

The model of Eq. (6.20) requires some means of calculating $\nu_{\mathrm{t}}$. Kinetic theory gives a quantitative prediction of $\nu$ in Eq. (6.19) which led Boussinesq to hypothesise that $\nu_{\mathrm{t}} \propto u\,l$, where $u$ and $l$ are a representative speed and length, respectively, with the speed $u$ relating to $\mathbf{u}'$ due to turbulence.

The turbulent viscosity can also be expressed as $\nu_{\mathrm{t}} = u_{\mathrm{m}}\,l_{\mathrm{m}}$ by absorbing the constant of proportionality within a characteristic speed $u_{\mathrm{m}}$ and a *mixing length* $l_{\mathrm{m}}$, discussed in Sec. 6.12.

Eddy viscosity and mixing length are useful concepts in turbulence modelling. However, it should be recognised that there are limitations in the analogy with kinetic theory, *e.g.*:

- momentum is exchanged between submicroscopic particles through intermittent, discrete collisions, compared to the continuous interaction between eddies;

- the magnitude of random particle motions is generally equal in all directions, whereas the level of turbulent fluctuations can sometimes vary significantly with direction.

## 6.12 Mixing length

By analogy with kinetic theory, the eddy viscosity $\nu_t$ can be expressed in terms of a characteristic speed $u_m$ and length $l_m$. Prandtl produced the following model for the mixing length $l_m$:†

$$\nu_t = l_m^2 \left| \frac{\partial u_x}{\partial y} \right| . \tag{6.21}$$

The mixing length Eq. (6.21) is derived from the kinetic view of viscosity in Sec. 6.10. Turbulent fluctuations in the $y$-direction, $u'_y$, carry fluid to the $y = 0$ plane from a distance $\pm l$.

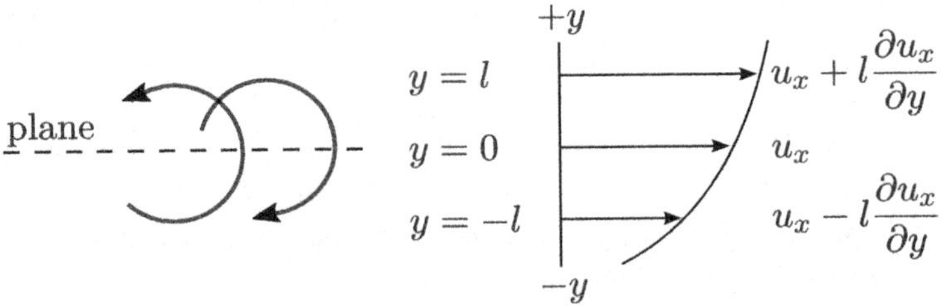


At $y = 0$, the turbulent fluctuations in the $x$-direction, $u'_x$, correspond to the range of velocities at $y = \pm l$ such that

$$|u'_x| \sim l \, |\partial u_x / \partial y| .$$

For a positive $\partial u_x/\partial y$, a positive $u'_x$ at $y = 0$ corresponds to a negative $u'_y$, and vice versa. The Reynolds stress $\tau_{yx}^t$ can then be constructed by defining a mixing length $l_m$ to replace $l$, absorbing the constant of proportionality, to give

$$\tau_{yx}^t = -\rho \overline{u'_x u'_y} = \rho l_m^2 \left| \frac{\partial u_x}{\partial y} \right| \frac{\partial u_x}{\partial y} . \tag{6.22}$$

From $\tau_{yx}^t = \rho \nu_t \, \partial u_x / \partial y$, we arrive at $\nu_t$ in Eq. (6.21).

---

†Ludwig Prandtl, *Bericht uber Untersuchungen zur ausgebildeten Turbulenz*, 1925.

The mixing length Eq. (6.21) is only effective as a turbulence model to calculate $\nu_\mathrm{t}$ when the flow is simple enough that $l_\mathrm{m}$ can be chosen appropriately.

Such an example is high Re, fully-developed flow through a pipe of radius $R$. The mixing length $l_\mathrm{m}$, calculated from measured velocity profiles, follows a polynomial function of distance $y$ from the wall, given by[†]

$$\frac{l_\mathrm{m}}{R} = 0.14 - 0.08\left(1 - \frac{y}{R}\right)^2 - 0.06\left(1 - \frac{y}{R}\right)^4. \tag{6.23}$$

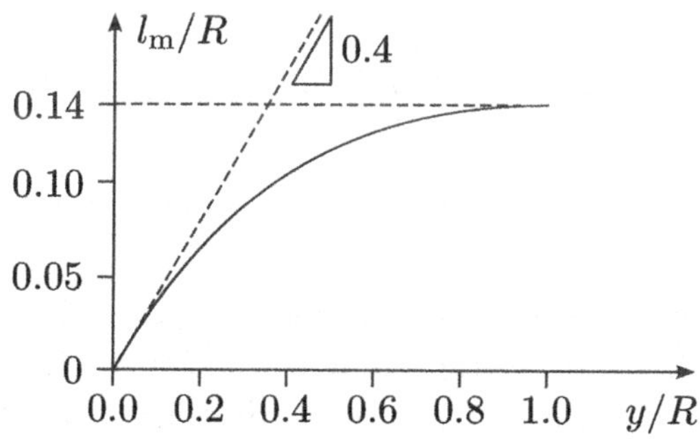


Notably, close to the wall, *e.g.* $y/R < 0.1$, the mixing length increases linearly according to

$$l_\mathrm{m} = \kappa y, \tag{6.24}$$

where $\kappa \approx 0.4$ is Kármán's constant.

At the centre of the pipe, $l_\mathrm{m} = 0.14R = 0.07D$ where $D$ is pipe diameter. Estimates of $l_\mathrm{m} = 0.07L - 0.09L$ are commonly cited for other simple examples, *e.g.* mixing layer, jet, flat plate boundary layer, *etc.*, where $L$ is the characteristic length of the problem (radius in the case of a jet).

[†]Johann Nikuradse, *Strömungsgesetze in rauhen Rohren*, 1933.

## 6.13 Turbulent kinetic energy

The Reynolds stress $\boldsymbol{\sigma}^{\mathrm{t}}$ in the ensemble-averaged momentum conservation Eq. (6.15) can be decomposed into viscous and pressure components. The turbulent pressure term, $(2/3)\rho k$, is commonly subsumed within pressure $p$, to give

$$\frac{\partial \rho \mathbf{u}}{\partial t} + \nabla \bullet \rho \mathbf{u}\mathbf{u} = \nabla \bullet \left(\boldsymbol{\tau} + \boldsymbol{\tau}^{\mathrm{t}}\right) - \nabla p + \rho \mathbf{b}. \tag{6.25}$$

The viscous stresses can be combined assuming $\boldsymbol{\tau}$ is modelled as a Newtonian fluid Eq. (2.41) and $\boldsymbol{\tau}^{\mathrm{t}}$ by the eddy viscosity model Eq. (6.20) to give

$$\frac{\partial \rho \mathbf{u}}{\partial t} + \nabla \bullet \rho \mathbf{u}\mathbf{u} = 2\nabla \bullet \rho \nu_{\mathrm{eff}} \,\mathrm{dev}\, \mathbf{D} - \nabla p + \rho \mathbf{b}, \tag{6.26}$$

where the *effective* kinematic viscosity

$$\nu_{\mathrm{eff}} = \nu + \nu_{\mathrm{t}}. \tag{6.27}$$

The effective viscosity represents momentum diffusion from the combined molecular and turbulent motions. The properties due to molecular motion are often described as *laminar*, *e.g.* laminar viscosity $\nu$.

Averaging the momentum equation and introducing the eddy viscosity model creates one additional unknown $\nu_{\mathrm{t}}$. Additional models are required for $\nu_{\mathrm{t}}$ to close the system of equations.

By considering $\nu_{\mathrm{t}} = u_{\mathrm{m}} l_{\mathrm{m}}$ from Sec. 6.11, the model for $\nu_{\mathrm{t}}$ is typically decomposed into components representing velocity and length scales, $u_{\mathrm{m}}$ and $l_{\mathrm{m}}$ respectively.

The scale of $u_{\mathrm{m}}$ corresponds to turbulent fluctuations $\mathbf{u}'$, so it is reasonable to assume that $u_{\mathrm{m}} \propto k^{1/2}$. Since the field $k$ is representative of the $u_{\mathrm{m}}$ component of $\nu_{\mathrm{t}}$, it is commonly adopted within turbulence models based on $\nu_{\mathrm{t}}$. Also, it is well captured by a suitable conservation equation.

## Transport of turbulent kinetic energy

Conservation of turbulent kinetic energy $k$ can be written as:

$$\boxed{\rho\frac{Dk}{Dt} = \nabla\bullet(\rho D_k\nabla k) + \rho G - \frac{2}{3}\rho k(\nabla\bullet\mathbf{u}) - \rho\varepsilon} \tag{6.28}$$

where the *turbulence generation* $G$ is

$$G = 2\nu_{\mathrm{t}}\,\mathrm{dev}\,\mathbf{D}:\nabla\mathbf{u}, \tag{6.29}$$

and $D_k$ is the effective diffusivity for $k$. The equation is derived in a manner similar to Eq. (2.56) for conservation of specific internal energy $e$, by (ensemble) averaging the separate energy contributions from $\overline{\mathbf{u}}$ and $\mathbf{u}'$. While $e$ includes the kinetic energy of molecular motion, $k$ is the equivalent for eddy motion.

In Eq. (2.56), energy from bulk motion is passed to the sub-microscopic scale as *heat* by $\boldsymbol{\sigma}:\nabla\mathbf{u}$. In Eq. (6.28), it is converted to *turbulent energy* by $\boldsymbol{\sigma}^{\mathrm{t}}:\nabla\mathbf{u}$ using Boussinesq's $\boldsymbol{\sigma}^{\mathrm{t}}$ from Eq. (6.20). The shear component $\boldsymbol{\tau}^{\mathrm{t}}$ provides the non-recoverable $\rho G$ in Eq. (6.28) and the second ($k$) term yields $^2/_3\rho k(\nabla\bullet\mathbf{u})$.

While $\rho G$ transfers kinetic energy from the bulk flow to $k$, the final term $\rho\varepsilon$ transfers $k$ on to $e$ as dissipated heat. Here, $\varepsilon$ is the *turbulent dissipation rate* per unit mass from in Sec. 6.6 which, from an ensemble-averaged derivation of Eq. (6.28), is

$$\varepsilon = \nu\overline{\nabla\mathbf{u}':\nabla\mathbf{u}'}. \tag{6.30}$$

Finally, diffusion of $k$ is represented by $\nabla\bullet(\rho D_k\nabla k)$, where $D_k = \nu + \nu_{\mathrm{t}}/\sigma_k$. This represents diffusion by both molecular and turbulent motions and interactions, including an adjustable coefficient $\sigma_k$ which is usually set to 1.

The $k$ equation goes part of the way to closing our system of equations. However, it introduces an *additional unknown*, $\varepsilon$, and the model for $\nu_{\mathrm{t}}$ still *requires length scale* $l_{\mathrm{m}}$.

## 6.14 Turbulent dissipation rate

A complete model for $\nu_t$ is still required to solve the ensemble-averaged momentum equation, *e.g.* Eq. (6.26). The discussion in Sec. 6.11 indicates $\nu_t$ is the product of $u_m$ and $l_m$, requiring two models to represent each scale. Since $u_m \propto k^{1/2}$, $k$ can represent the speed scale, modelled by Eq. (6.28).

The turbulent dissipation rate $\varepsilon$ matches the rate of transfer of kinetic energy down the energy cascade $\sim u^3/l$, as discussed in Sec. 6.6. This applies to all turbulent scales including the larger mixing length scales, so $\varepsilon \propto k^{3/2}/l_m$. Substituting $u_m$ and $l_m$ expressions into $\nu_t = u_m\, l_m$ yields turbulent viscosity

$$\nu_t = c_\mu \frac{k^2}{\varepsilon}, \tag{6.31}$$

where $c_\mu$ is a constant. From empirical data, $c_\mu = 0.09$, except within the viscous and buffer layers close to a wall, see Sec. 7.4.

### Transport of turbulent dissipation rate

A model is now required for $\varepsilon$, both to calculate $\nu_t$ by Eq. (6.31) and to provide the remaining unknown in Eq. (6.28). The model can be provided by a transport equation for $\varepsilon$

$$\boxed{\begin{aligned}\rho\frac{D\varepsilon}{Dt} &= \nabla \cdot (\rho D_\varepsilon \nabla\varepsilon) + c_1 \rho G \frac{\varepsilon}{k} \\ &\quad - \frac{2}{3} c_1 \rho \varepsilon (\nabla \cdot \mathbf{u}) - c_2 \rho \frac{\varepsilon^2}{k}\end{aligned}} \tag{6.32}$$

where $D_\varepsilon = \nu + \nu_t/\sigma_\varepsilon$ is the combined molecular and turbulent diffusion with an adjustable coefficient $\sigma_\varepsilon$, usually set to 1.3. The remaining coefficients $c_1 = 1.44$ and $c_2 = 1.92$ are tuned to capture the behaviour of a range of flows.

The $\varepsilon$-*equation*, Eq. (6.32), can be derived in terms of statistical properties, replacing high-order terms in $\mathbf{u}'$ by models with coefficients $c_1$, $c_2$.[†]

Alternatively, it can be obtained by multiplying the principal variable, $k$ or $\varepsilon$, in each term of Eq. (6.28) by $\varepsilon/k$ and introducing coefficients $\sigma_\varepsilon$, $c_1$ and $c_2$.

The $c_1 \rho G \varepsilon/k$ term in Eq. (6.32) causes $\varepsilon$ to increase with $G$. This is logical since the generated turbulence moves down the energy cascade, so ultimately affects the rate of dissipation.

The $c_2 \rho \varepsilon^2/k$ term is justified by considering the *free decay of turbulence.* If the fluid stops moving ($\mathbf{u} = 0$) and turbulence is no longer generated ($G = 0$), then (assuming constant $\rho$ and ignoring diffusion) Eq. (6.28) and Eq. (6.32) reduce to

$$dk/dt = -\varepsilon \quad \text{and} \quad d\varepsilon/dt = -c_2\, \varepsilon^2/k, \tag{6.33}$$

respectively. Integrating the combined equations yields $\varepsilon/\varepsilon_0 = (k/k_0)^{c_2}$ where the "0" subscript indicates initial values.

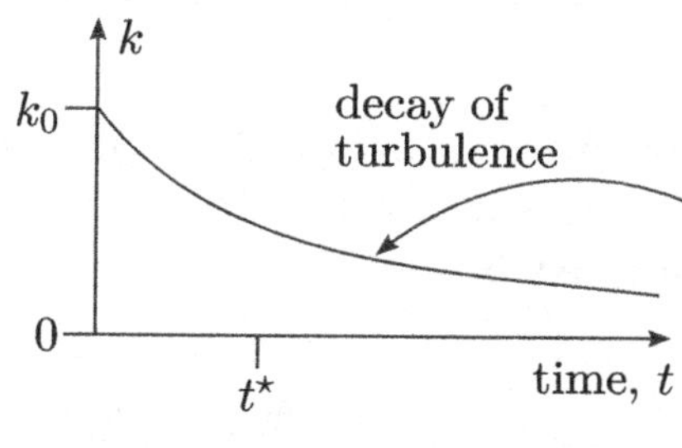


$$\int_{k_0}^{k} (k/k_0)^{-c_2}\, dk = -\int_0^t \varepsilon_0\, dt$$

$$\Rightarrow k = k_0\,(1 + t/t^\star)^{-n}$$

Further integration yields a decay in $k$ over time to the power $n = 1/(c_2 - 1) = 1.09$ and a timescale $t^\star = n k_0/\varepsilon_0$, which is a reasonable approximation to real behaviour.

---

[†]B. I. Davydov, *Statistical dynamics of an incompressible turbulent fluid, Dokl. Akad. Nauk SSSR* **136**, 1961.

## 6.15 Summary of turbulence

- Turbulence is a spatially complex distribution of vorticity which advects itself in a chaotic manner, Sec. 6.2.
- Vorticity is mostly generated at solid boundaries and evolves by advection, diffusion and stretching/compression, Sec. 6.3.
- Boundary layers are thus the main source of vorticity for turbulence, Sec. 6.4.
- Turbulence occurs when instabilities cause the vorticity to become chaotic, characterised by Reynolds Number, Sec. 6.1.

### Scales of turbulence

- Length and time scales of turbulent eddies initially relate to the characteristic scales of the generated vortices, *e.g.* by shedding, Sec. 6.5.
- The eddies become progressively smaller until they reach a size where the dissipation of their kinetic energy is significant, Sec. 6.6.
- The process of eddy breakup involves a cascade of kinetic energy from larger to smaller eddies in which eddy structures become increasingly isotropic, Sec. 6.7.

### Simulating turbulence

- The Kolmogorov microscales represent the smallest scales of turbulence at which eddies die out by viscous dissipation.
- They are surprisingly small, *e.g.* a factor of $\mathrm{Re}^{3/4}$ separates the length scales of the largest and smallest eddies.

- Usually, it is prohibitively costly to capture the smallest scales in a CFD simulation, Sec. 6.8.
- Large-eddy simulation captures the large eddies, while using modelling for the unresolved scales.
- Reynolds-averaged simulations provide faster solutions by solving equations for averaged field variables, Sec. 6.9.
- The averaged momentum equation includes the Reynolds stress which requires modelling.

## Turbulence modelling concepts

- Momentum exchange at the molecular scale is characterised by viscosity and kinetic theory provides good quantitative predictions of viscosity for gases, Sec. 6.10.
- Comparing the random motion of eddies in a turbulent fluid with particles at a molecular scale, Boussinesq introduced the concept of eddy viscosity, Sec. 6.11.
- Eddy viscosity is not a fluid property but is proportional to a characteristic speed and length scale of the turbulence; models are required which represent each of these scales.
- The speed scale is represented by turbulent kinetic energy $k$, described by a transport equation Eq. (6.28), Sec. 6.13.
- The $k$-equation includes a term for its rate of dissipation $\varepsilon$; a transport equation for $\varepsilon$, *e.g.* Eq. (6.32), provides a model for that term — which also represents the length scale of turbulence, Sec. 6.14.
- The concept of mixing length to characterise the length scale of turbulence provides additional modelling, particularly for the near-wall region, Sec. 6.12.

# Chapter 7

# Reynolds-Averaged Turbulence Modelling

> *"I get the news I need on the weather report. Oh, I can gather all the news I need on the weather report."*

Simon & Garfunkel, *The Only Living Boy in New York* (1970).

For CFD to be useful, it must calculate the flow, forces, heat transfer, *etc.* with sufficient accuracy. The accuracy of any calculation is always compromised by various approximations.

First, there are approximations from the numerical methods that are employed. Discretisation schemes, described in Chapter 3, affect accuracy which can be improved by reducing the size of cells and time step, at the cost of increased solution times.

The algorithms and solvers, described in Chapter 5, are a further source of numerical approximation. Linear solvers are generally iterative with tolerances that balance computational cost against accuracy of the solution. Algorithms may include loops with a fixed number of iterations that limits the overall solution convergence.

Boundary conditions, described in Chapter 4, are applied as part of the configuration of the CFD simulation. Invariably there is some level of approximation in the way the conditions represent the problem of interest. The approximation is physical, not numerical, and includes everyday examples such as:

- setting a fixed value condition with a uniform value, *e.g.* on velocity at an inlet;
- applying a zero gradient condition, *e.g.* to pressure at a wall boundary;
- imposing the direction of inflow, *e.g.* with the inlet-outlet-velocity condition, Sec. 4.15.

Beyond the numerical methods and boundary conditions, are the models that describe fluid dynamics. The governing equations and basic models in Chapter 2 are accurate for laminar flows with modest speeds and heat transfer. However, as the flow physics becomes more complex, the models become increasingly approximate.

It is clear from Chapter 6 that turbulence is highly complex, so turbulence modelling is often a major source of error in a CFD solution. It exemplifies the observation of statistician George Box that *all models are wrong.*[†] He advises that we should seek "an economical description of natural phenomena" and "simple but evocative models".

It is with that advice we present some of the Reynolds-averaged turbulence models commonly used in CFD. The models take the form of transport equations for turbulence fields, typically turbulent kinetic energy and dissipation rate.

The models are accompanied by advice on specification of the inlet and initial conditions for the fields. Particular attention is given to the modelling of the flow within turbulent boundary layers at wall boundaries.

---

[†] George Box, Science and statistics, 1976.

## 7.1 The $k$–$\varepsilon$ turbulence model

The $k$–$\varepsilon$ **model** (k-epsilon) was the first turbulence model to be widely adopted for a variety of flows in CFD.† It is one of a family of *two equation* models which solve two transport equations, one usually for $k$ and one for another variable, often a dissipation rate. These models are the industry standard for CFD.

The $k$–$\varepsilon$ model combines the transport equations for $k$ and $\varepsilon$, Eq. (6.28) and Eq. (6.32) respectively, which are reproduced below assuming $\rho =$ constant and replacing material derivatives in conservative form as in Eq. (2.26).

$$\frac{\partial k}{\partial t} + \nabla \bullet (\mathbf{u}k) - \nabla \bullet (D_k \nabla k) = G - \frac{2}{3} k (\nabla \bullet \mathbf{u}) - \varepsilon \qquad (7.1)$$

$$\begin{aligned} &\frac{\partial \varepsilon}{\partial t} + \nabla \bullet (\mathbf{u}\varepsilon) - \nabla \bullet (D_\varepsilon \nabla \varepsilon) = \\ &\qquad c_1 G \frac{\varepsilon}{k} - \frac{2}{3} c_1 \varepsilon (\nabla \bullet \mathbf{u}) - c_2 \frac{\varepsilon^2}{k} \end{aligned} \qquad (7.2)$$

The standard model coefficients are

$$\begin{aligned} &D_k = \nu + \nu_t / \sigma_k, \quad D_\varepsilon = \nu + \nu_t / \sigma_\varepsilon, \\ &c_\mu = 0.09, \quad c_1 = 1.44, \quad c_2 = 1.92, \\ &\sigma_k = 1.0, \quad \sigma_\varepsilon = 1.3. \end{aligned} \qquad (7.3)$$

The choice of coefficients in Eq. (7.3) originates from a large series of validation simulations of free shear flows.‡

†William Jones and Brian Launder, *The prediction of laminarization with a two-equation model of turbulence*, 1972.

‡Brian Launder, Alan Morse, Wolfgang Rodi and Brian Spalding, *Prediction of free shear flows — A comparison of the performance of six turbulence models*, NASA Conference on Free Shear Flows, Langley, 1972.

The model can be deployed in a steady flow solution, setting the local time derivatives to zero, *e.g.* $\partial k/\partial t \equiv 0$. It can also form part of a transient solution to capture unsteady features such as vortex shedding.

The model uses a momentum equation in ensemble-averaged form, *e.g.* Eq. (6.26), using $\nu_{\text{eff}}$ calculated from updated $\nu$ and $\nu_{\text{t}}$ according to Eq. (6.27).

The model equations are added to the end of the main loop in the steady and transient algorithms from Sec. 5.12 or Sec. 5.21, respectively. The figure below shows the additions to the transient algorithm, starting from the momentum corrector step.

| | | ***PIMPLE loop*** |
|---|---|---|
| **Momentum corrector** | $\mathbf{u} := \dfrac{\mathbf{H}(\mathbf{u})}{A} - \dfrac{1}{A}\nabla p$ | |
| **Flux corrector** | calculate $\phi_{\text{f}} := \ldots$ | |
| **Update $\nu$** | calculate $\nu := \ldots$ | |
| **Sources** | calculate $G$ and $\nabla \cdot \mathbf{u}$ | |
| **$\varepsilon$ equation** | solve $\partial\varepsilon/\partial t \ldots$ | |
| **$k$ equation** | solve $\partial k/\partial t \ldots$ | |
| **Update $\nu_{\text{t}}$** | calculate $\nu_{\text{t}} = c_\mu\, k^2/\varepsilon$ | |
| | $p, \mathbf{u}, \phi_{\text{f}}, T, k, \varepsilon, \nu, \nu_{\text{t}}$ | |
| | **ENDWHILE** | |

Following the momentum and flux correctors, $\nu$ can be updated, *e.g.* for non-Newtonian models where $\nu = f(\mathbf{D}) = f(\nabla\mathbf{u})$. $G$ and $\nabla \cdot \mathbf{u}$ are calculated and stored for source terms in both the $k$ and $\varepsilon$ equations. Those equations are then solved, followed by an update to $\nu_{\text{t}}$, and subsequently $\nu_{\text{eff}}$.

## 7.2 Initialisation of the $k$–$\varepsilon$ model

Initial values and boundary conditions must be specified for $k$ and $\varepsilon$ to solve their respective transport equations. The ideal specification of boundary conditions for $k$ and $\varepsilon$ follows those for $T$ described in Sec. 4.3.

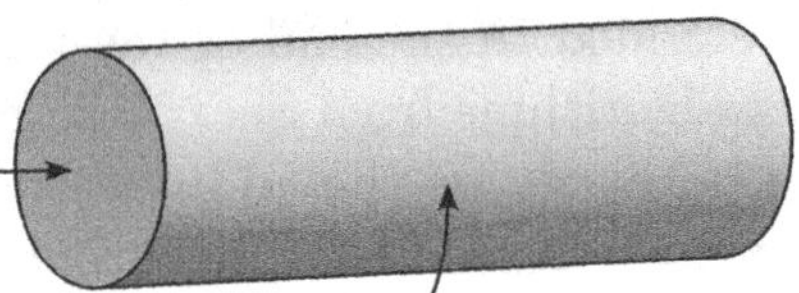


Turbulence fields require: a fixed value condition at inlets; zero gradient or inlet-outlet at outlets; and, a more complex specification at solid walls, introduced in Sec. 7.7.

Inlet values $k_{\text{in}}$ and $\varepsilon_{\text{in}}$ must therefore be specified. There may be industry standards, published recommendations or measured data to help select these values for the specific problem being simulated.

But more often than not, $k_{\text{in}}$ and $\varepsilon_{\text{in}}$ must be estimated. Inlet and initial estimates of $\varepsilon_{\text{in}}$ are usually based on a Prandtl mixing length $l_{\text{m}}$ from the expression

$$\varepsilon_{\text{in}} = c_\mu^{3/4} \frac{k^{3/2}}{l_{\text{m}}}. \tag{7.4}$$

This relation can be derived by considering a planar boundary layer. In the "log law" region (see Sec. 7.7) $\varepsilon \approx G$ which can be combined with the mixing length Eq. (6.21) to give

$$\varepsilon \approx G = \nu_\mathrm{t} \left( \frac{\partial u_x}{\partial y} \right)^2 = \nu_\mathrm{t}^3 / l_\mathrm{m}^4. \tag{7.5}$$

Substituting $\nu_\mathrm{t}$ from Eq. (6.31) yields Eq. (7.4).

A value for $l_\mathrm{m}$ must then be specified in order to calculate inlet and initial values of $\varepsilon_\mathrm{in}$ from Eq. (7.4). Procedures to estimate $l_\mathrm{m}$ are described in Sec. 7.3.

Inlet and initial estimates for $k_\mathrm{in}$ can be calculated by

$$k_\mathrm{in} = \frac{3}{2} \left( |\mathbf{u}| I \right)^2, \tag{7.6}$$

from a specified *turbulent intensity* $I = \mathbf{u}'_\mathrm{RMS} / |\mathbf{u}|$, the ratio of the root-mean-square (RMS) of turbulent fluctuations $\mathbf{u}'_\mathrm{RMS}$ to the mean flow speed $|\mathbf{u}|$. The expression is derived from the definition $\mathbf{u}'_\mathrm{RMS} = (\overline{\mathbf{u}'^2}/3)^{1/2} = (2k/3)^{1/2}$.

A value for $I$ must then be specified in order to calculate the inlet and initial values of $k_\mathrm{in}$ from Eq. (7.6). Procedures to estimate $I$ are also described in Sec. 7.3. The values of $k_\mathrm{in}$ and $\varepsilon_\mathrm{in}$ at inlet boundaries influence the solution throughout the CFD simulation, so should be estimated as accurately as possible.

The *accuracy* of the initial (internal) values is not so critical, since they do not influence the solution beyond a short period at the beginning of a simulation.

Initial values can, however, affect *stability* during the early steps of a CFD simulation. The flow boundary conditions generally cause sudden impulses which can generate large forces, causing fluctuations in the solution of $p$. Higher $\nu_\mathrm{t}$, based on initial $k$ and $\varepsilon$ values, tends to cause larger fluctuations, which may make the solution of the $p$-equation unstable.

## 7.3 Inlet turbulence

Expressions are presented in Sec. 7.2 to estimate inlet and initial values of $k_{\mathrm{in}}$ and $\varepsilon_{\mathrm{in}}$. They include parameters $I$ and $l_{\mathrm{m}}$ which must themselves be estimated sufficiently accurately to calculate $k_{\mathrm{in}}$ and $\varepsilon_{\mathrm{in}}$ reliably.

The values of $I$ and $l_{\mathrm{m}}$ at domain inlets depend on the flow conditions upstream of the inlet. The figure below shows typical ranges of intensity $I$ for different upstream flow conditions.

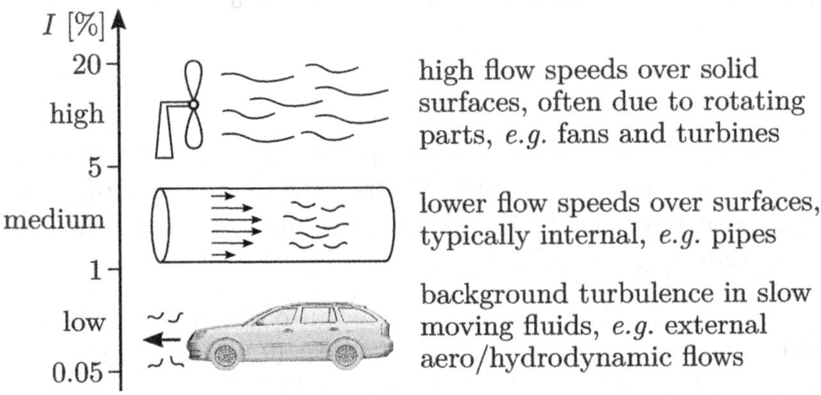


A medium intensity $1\% < I < 5\%$ is most commonly specified in CFD problems, in particular for internal flows. For these flows, $I$ can be calculated from a power-law function of Re, fitted to measurements at the central axis in fully developed flow along a smooth-wall pipe, according to[†]

$$I = 0.055\mathrm{Re}^{-0.041}. \tag{7.7}$$

An estimate of $l_{\mathrm{m}} = 0.07D$ at the centre axis of a pipe, see Sec. 6.12, can be used in conjunction with $I$ from of Eq. (7.7).

[†] Nils Basse, *Turbulence intensity and the friction factor for smooth- and rough-wall pipe flow*, 2017.

For ducts and channels of non-circular cross-section, $D$ can be calculated by $D = 4 \times A/P$, where $A$ is the cross-sectional area and $P$ is the perimeter length. For a partially filled pipe or duct, $P$ corresponds to the wetted region where the fluid is in contact with the boundary.

For wall-bounded flows with a boundary layer of thickness $\delta$, an estimate of $l_{\mathrm{m}} = \kappa\delta \approx 0.4\delta$ is often used. This relation (see also Sec. 6.12) requires $\delta$ to be estimated, *e.g.* from the $\delta/x$ expression for a turbulent layer at the end of Sec. 6.4.

## Verifying turbulent viscosity

Combining Eq. (7.4), Eq. (7.6) and Eq. (6.31) gives the following expression for $\nu_{\mathrm{t}}$ in terms of length $l_{\mathrm{m}}$ and velocity $|\mathbf{u}|I$ scales:

$$\nu_{\mathrm{t}} = c_\mu \frac{k_{\mathrm{in}}^2}{\varepsilon_{\mathrm{in}}} = \left(\frac{3}{2}\right)^{\frac{1}{2}} c_\mu^{1/4}\, |\mathbf{u}|I\, l_{\mathrm{m}} \approx 0.67\, |\mathbf{u}|I\, l_{\mathrm{m}}. \tag{7.8}$$

Values of $\nu_{\mathrm{t}}$ need to be realistic. Realistic values usually fall within the range of molecular viscosities $\nu$ for common fluids at standard temperature shown below.

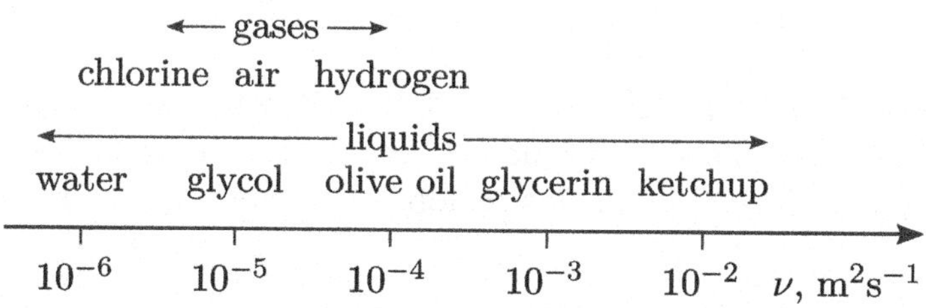


The range is presented in terms of *kinematic* viscosity $\nu$ which governs the *rate* of momentum *diffusion*, *e.g.* the rate of growth of boundary layers. By contrast, *forces* are governed by dynamic viscosity $\mu = \rho\nu$, which make liquids "feel" more viscous.

## 7.4 Turbulent boundary layers

At solid walls, the tangential flow speed $u_x$ increases rapidly across a thin boundary layer, as discussed in Sec. 6.4. At high Re, the velocity profile has a universal character shown below.

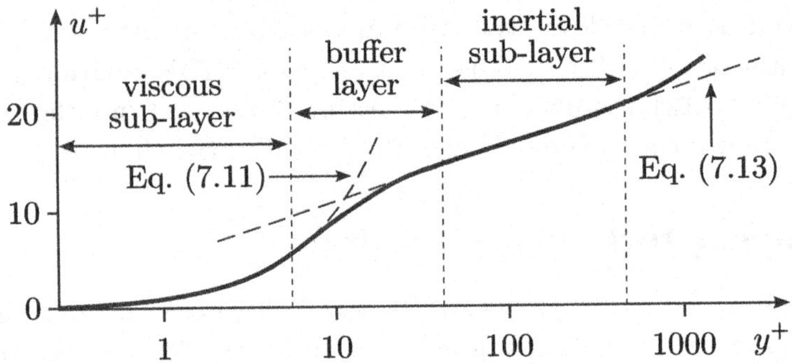


The profile compares measured data† in terms of a dimensionless velocity $u^+$ and distance to the wall $y^+$, given by

$$u^+ = u_x/u_\tau \quad \text{and} \quad y^+ = u_\tau y/\nu. \tag{7.9}$$

Both parameters in Eq. (7.9) are based on a *friction velocity* $u_\tau$ which is related to the wall shear stress $\tau_\mathrm{w}$ by

$$\tau_\mathrm{w} = \rho u_\tau^2. \tag{7.10}$$

At the wall $\mathbf{u}' = 0$. Close to the wall, $\mathbf{u}'$ is suppressed, creating a region where flow is laminar $y^+ < 5$, known as the *viscous sub-layer* . The profile in this region is described by the relation

$$\boxed{u^+ = y^+} \tag{7.11}$$

Turbulence becomes significant through the *buffer layer* which describes the region $5 < y^+ < 40$. Van Driest provides a model

†Joseph Kestin and Peter Richardson, *Heat transfer across turbulent, incompressible boundary layers*, 1963.

for the increase in mixing length through this region, by[†]

$$l_{\mathrm{m}} = \kappa y \left[1 - \exp(-y^+/26)\right]. \tag{7.12}$$

Finally, in the *inertial sub-layer* for $y^+ > 40$, flow is turbulent and the velocity profile is described by the *logarithmic law of the wall*, often abbreviated to simply the *log law*, according to[‡]

$$\boxed{u^+ = \frac{1}{\kappa}\ln(y^+) + B} \tag{7.13}$$

The equation includes Kármán's constant $\kappa = 0.41$ and constant $B$. For a smooth wall, $B = 5.0$[§] – 5.5 is commonly used. Both Eq. (7.11) and Eq. (7.13) can be derived assuming a constant shear stress across the profile, equating to $\tau_{\mathrm{w}}$ at the wall. In the viscous sub-layer, the shear stress is laminar, so

$$\frac{\tau_{\mathrm{w}}}{\rho} = u_\tau^2 = \nu\frac{\partial u_x}{\partial y}. \tag{7.14}$$

This equation integrates with a zero constant of integration to give $u_\tau^2 y = \nu u_x$, from which Eq. (7.11) is derived. In the inertial sub-layer, the shear stress is turbulent (laminar is negligible), so

$$\frac{\tau_{\mathrm{w}}}{\rho} = u_\tau^2 = \nu_{\mathrm{t}}\frac{\partial u_x}{\partial y} = l_{\mathrm{m}}^2\left|\frac{\partial u_x}{\partial y}\right|\frac{\partial u_x}{\partial y}. \tag{7.15}$$

Assuming $l_{\mathrm{m}} = \kappa y$ gives $u_\tau = \kappa y\, \partial u_x/\partial y = \kappa y^+\, \partial u_x/\partial y^+$, which integrates to yield Eq. (7.13). In the inertial sub-layer, $\varepsilon \approx G$ as described in Eq. (7.5), which combines with Eq. (7.15) and Eq. (6.31) to give

$$u_\tau = (\nu_{\mathrm{t}}\varepsilon)^{1/4} = c_\mu^{1/4} k^{1/2}. \tag{7.16}$$

[†] Edward van Driest, *On turbulent flow near a wall*, 1956.

[‡] Alternatively written $u^+ = \ln(Ey^+)/\kappa$, where $E = \exp(\kappa B)$.

[§] Hermann Schlichting and Klaus Gersten, *Boundary-layer theory*, 2017.

## 7.5 Wall functions

CFD simulations may be used to calculate the forces on solid bodies exerted by the fluid, *e.g.* in aerodynamics. The wall shear stress is then calculated according to $\tau_{\mathrm{w}} = \rho\nu\, \partial u_x/\partial y$. With turbulent boundary layers, the $\partial u_x/\partial y$ calculation requires cells with very small lengths normal to the wall to be accurate. The resulting mesh is inevitably large, which carries a high computational cost. The problem for CFD is how to calculate $\tau_{\mathrm{w}}$ with sufficient accuracy, but at an affordable cost.

*Wall functions* provide a solution to this problem by exploiting the universal character of the velocity distribution described in Sec. 7.4. They use the law of the wall Eq. (7.13) as a model to provide a reasonable prediction of $\tau_{\mathrm{w}}$ from a relatively inaccurate $\partial u_x/\partial y$ calculation at the wall.

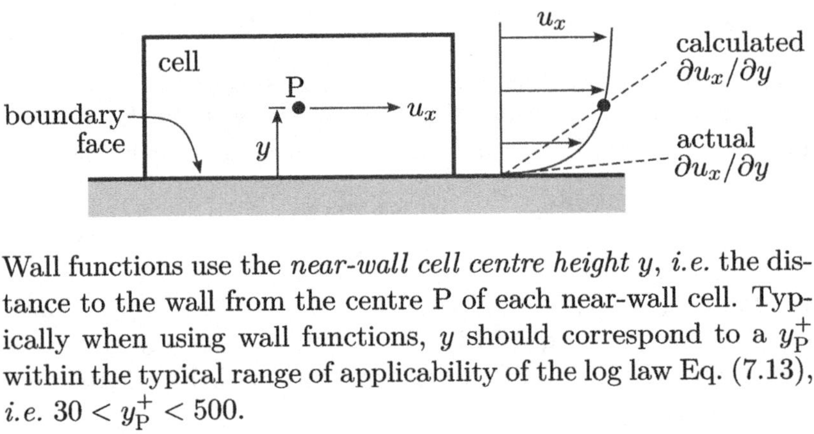


Wall functions use the *near-wall cell centre height* $y$, *i.e.* the distance to the wall from the centre P of each near-wall cell. Typically when using wall functions, $y$ should correspond to a $y^+_{\mathrm{P}}$ within the typical range of applicability of the log law Eq. (7.13), *i.e.* $30 < y^+_{\mathrm{P}} < 500$.

With such a mesh, the calculated $\partial u_x/\partial y$ is then significantly lower than its true value. Wall function models compensate for the resulting error in the prediction of $\tau_{\mathrm{w}}$ by increasing viscosity at the wall. The increase is applied to $\nu_{\mathrm{t}}$ at the wall patch faces, which would otherwise be $\nu_{\mathrm{t}} = 0$, corresponding to $\mathbf{u}' = 0$.

## Standard wall function

The **standard** wall function for a "smooth" wall calculates $\nu_\mathrm{t}$ for each patch face based on the near-wall $y^+_\mathrm{P}$. No adjustment is made to $\nu_\mathrm{t}$ when $y^+_\mathrm{P}$ corresponds to the viscous sub-layer. When $y^+_\mathrm{P}$ corresponds to the inertial sub-layer, $\nu_\mathrm{t}$ is calculated by

$$\nu_\mathrm{t} = \nu \left( \frac{\kappa y^+_\mathrm{P}}{\ln(y^+_\mathrm{P}) + \kappa B} - 1 \right) \quad \text{for} \quad y^+_\mathrm{P} > y^+_\mathrm{tr}. \tag{7.17}$$

The condition ($y^+_\mathrm{P} > y^+_\mathrm{tr}$) that determines whether $y^+_\mathrm{P}$ lies within the inertial sub-layer corresponds to $y^+_\mathrm{tr}$ at the intersection of Eq. (7.11) and Eq. (7.13), calculated iteratively as

$$y^+_\mathrm{tr} = \ln(y^+_\mathrm{tr})/\kappa + B \approx 11. \tag{7.18}$$

$y^+_\mathrm{P}$ is calculated in each near-wall cell according to:

$$y^+_\mathrm{P} = c_\mu^{1/4} \left( \frac{k^{1/2} y}{\nu} \right)_\mathrm{P}. \tag{7.19}$$

This expression is derived from Eq. (7.9) for $y^+$ and Eq. (7.16). The subscript $(\ldots)_\mathrm{P}$ denotes all properties are evaluated at P.

The wall function Eq. (7.17) is derived from the notion that $\tau_\mathrm{w}$ is calculated numerically (assuming a stationary wall) by:

$$\frac{\tau_\mathrm{w}}{\rho} = \nu_\mathrm{eff} \frac{\partial u_x}{\partial y} = (\nu + \nu_\mathrm{t}) \frac{u_x}{y}. \tag{7.20}$$

At the same time, combining Eq. (7.9) and Eq. (7.10) gives

$$\frac{\tau_\mathrm{w}}{\rho} = \nu \frac{y^+}{u^+} \frac{u_x}{y}. \tag{7.21}$$

Comparing Eq. (7.20) and Eq. (7.21) gives $\nu_\mathrm{t} = \nu(y^+/u^+ - 1)$, which combines with Eq. (7.13) to yield Eq. (7.17).

## 7.6 Alternative wall functions

The standard wall function described in Sec. 7.5 uses a function for $\nu_\mathrm{t}$ that is discontinuous at $y^+ = y^+_\mathrm{tr}$, switching to $\nu_\mathrm{t} = 0$ for $y^+ \leq y^+_\mathrm{tr}$. A **continuous** wall function is available which evaluates $\nu_\mathrm{t}$ as $y^+ \to 0$ from a single equation describing the universal character of the velocity profile at high Re,†

$$y^+ = u^+ + \frac{1}{E}\left(\exp(\kappa u^+) - \sum_{n=0}^{3}\frac{(\kappa u^+)^n}{n!}\right), \tag{7.22}$$

where $E = \exp(\kappa B)$. The equation combines Eq. (7.11) and Eq. (7.13), "disabling" Eq. (7.13) at low $y^+$ by subtracting low order terms from a polynomial expansion of $\exp(\kappa u^+)$.

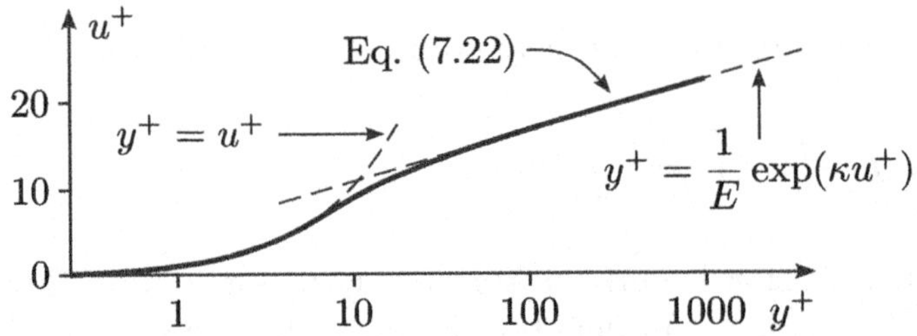


The wall function is applied by solving Eq. (7.22) for $u^+$ from $y^+$ calculated from Eq. (7.9) using the near-wall cell centre height $y$. The friction velocity is calculated by $u_\tau = u^+|\mathbf{u}_\mathrm{P}|$, where $\mathbf{u}_\mathrm{P}$ is the near-wall cell velocity. Finally, $\nu_\mathrm{t}$ on the wall patch is calculated from a numerical interpretation of Eq. (7.20),

$$\nu_\mathrm{t} = u_\tau^2/|\nabla_\mathbf{n}\mathbf{u}| - \nu, \tag{7.23}$$

where $\nabla_\mathbf{n}\mathbf{u}$ is the surface-normal velocity gradient. An iterative method is required to invert Eq. (7.22) and to accommodate other nonlinearities, *e.g.* $y^+$ is itself a function of $u_\tau$.

†Brian Spalding, *A single formula for the law of the wall*, 1961.

## Rough wall function

The standard wall function in Sec. 7.5 is applicable to smooth walls so does not account for surface roughness. Roughness is significant when the roughness "scale" $r$[†] becomes larger than the thickness of the viscous sub-layer.

At higher surface roughness, turbulent eddies are generated near the wall at a scale $\sim r$, rather than $\sim \nu/u_\tau$. The viscous effects become negligible, causing the non-dimensionalised distance to become $y^+ = y/r$ in the log law Eq. (7.13). To reflect this, Eq. (7.13) is modified to a form

$$u^+ = \frac{1}{\kappa}\ln(y^+) + B - \Delta u^+, \tag{7.24}$$

where $\Delta u^+$ a roughness function, dependent on $r^+ = u_\tau r/\nu$. An intuitive model for $\Delta u^+$ is[‡]

$$\Delta u^+ = \frac{1}{\kappa}\ln(1 + r^+). \tag{7.25}$$

This **rough** wall function Eq. (7.24) reduces to Eq. (7.13) using the conventional $y^+$ definition of Eq. (7.9) at $r^+ = 0$. As $r^+ \to \infty$, it reduces to Eq. (7.13) using $y^+ = y/r$.

It is open to interpretation how to determine $r$ from roughness measurements of a surface. The parameter is sometimes split into $r = C_\mathrm{s} r_\mathrm{s}$, where $r_\mathrm{s}$ is a measured *sand grain roughness height* and $C_\mathrm{s}$ is a coefficient that depends on the shape, consistency and packing of the roughness elements. Using that approach, values of $0.25 < C_\mathrm{s} < 0.50$ often yield a good match between Eq. (7.24) and measured data.

---

[†]'$k$' is often used to denote roughness, but we use '$r$' to avoid confusion with turbulent kinetic energy.

[‡]Cyril Colebrook, *Turbulent flow in pipes, with particular reference to the transitional region between smooth and rough wall laws*, 1939.

## 7.7 Turbulence near walls

The characteristic velocity distribution in turbulent boundary layers in Sec. 7.4, provided wall functions expressed as boundary conditions for $\nu_t$ in Sec. 7.5 and Sec. 7.6. Boundary conditions also need to be specified for turbulence fields at solid walls.

The turbulence generation $G$ influences the distribution of turbulence fields near a wall. At the wall, $G = 0$. In the inertial sub-layer, $G = u_\tau^3/\kappa y$ from Eq. (7.5) with: $\nu_t = u_\tau l_m$, obtained by combining Eq. (6.21) and Eq. (7.15); and, Eq. (6.24).

Since $G$ decreases with increasing $y$ in the inertial sub-layer, $G$ passes through a peak within the buffer layer (at $10 < y^+ < 15$).

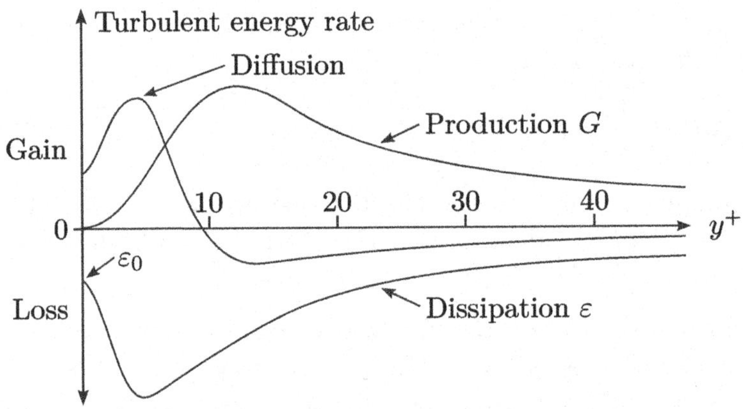


The peak in $G$ causes a similar peak in $k$, shown on the following page. To the left of the peak, turbulent energy is transported back towards the wall by diffusion $D_k\, \partial^2 k/\partial y^2$. The profile of dissipation $\varepsilon$ results from $G + D_k\, \partial^2 k/\partial y^2 - \varepsilon = 0$, obtained from Eq. (7.1). Very close to the wall ($y^+ < 5$), diffusion is predominately molecular, such that it non-zero at the wall. The dissipation $\varepsilon$ also has a non-zero value $\varepsilon_0$ at the wall.

## Wall functions and turbulence fields

When using a turbulence model, such as the $k$–$\varepsilon$ model described in Sec. 7.1, boundary conditions must be specified for $k$ and $\varepsilon$ at solid walls. The distribution of $k$, non-dimensionalised as $k^+ = k/u_\tau^2$, close to the wall is shown below.

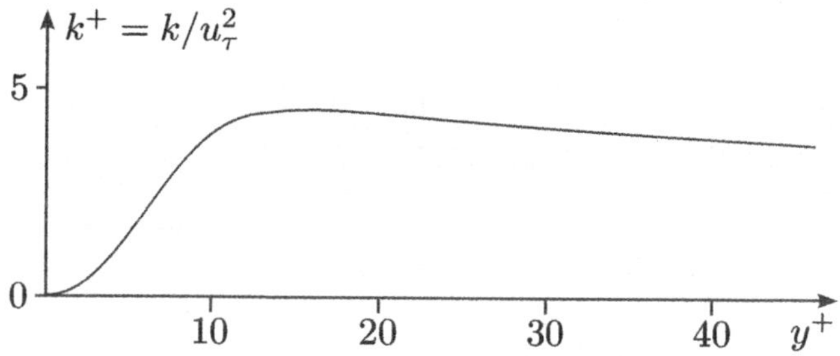


At the wall, $k = 0$ but it rises quickly to a peak at $10 < y^+ < 15$ before levelling off at $k^+ \approx 3.3$ as $y^+ \to \infty$.

With wall functions, the height of the centre of each near-wall cell should correspond to $y_\mathrm{P}^+$ within the range $30 < y_\mathrm{P}^+ < 500$. Viewed at that scale, the $k$ profile appears flat. For $\varepsilon$, there is no such simple profile shape. These observations lead to the boundary conditions for $k$ and $\varepsilon$ when using wall functions:

- **zero gradient** for $k$;
- **calculated near-wall cell value** for $\varepsilon$, according to:

$$\varepsilon = \begin{cases} c_\mu^{3/4} k_\mathrm{P}^{3/2} / \kappa y_\mathrm{P} & \text{for} \quad y_\mathrm{P}^+ > y_\mathrm{tr}^+; \\ 2\nu k_\mathrm{P} / y_\mathrm{P}^2 & \text{for} \quad y_\mathrm{P}^+ \le y_\mathrm{tr}^+. \end{cases} \tag{7.26}$$

The calculation uses $k_\mathrm{P}$ from the near-wall cell. The expression for $y_\mathrm{P}^+ > y_\mathrm{tr}^+$ is Eq. (7.4) with Eq. (6.24). The expression for $y_\mathrm{P}^+ \le y_\mathrm{tr}^+$ uses the asymptotic condition $y \to 0$ for $\varepsilon$ in Eq. (7.28).

## 7.8 Resolving the viscous sub-layer

Wall functions were introduced in Sec. 7.5 to avoid the need for a large mesh with cells small enough to resolve the boundary layer into the viscous sub-layer. They provide a reasonable prediction of $\tau_{\mathrm{w}}$ using the log law for the velocity distribution in the inertial sub-layer.

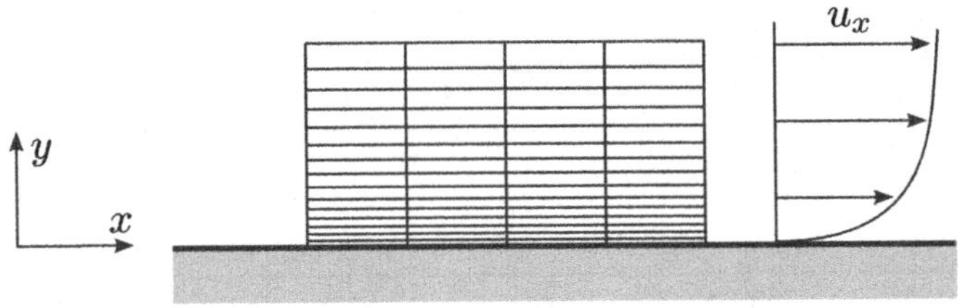


A CFD simulation may alternatively use a mesh with sufficiently thin cells to resolve the flow through the viscous sub-layer, *e.g.* with near-wall cell centre height corresponding to $y^+_{\mathrm{P}} = 1$, for a more accurate prediction of $\tau_{\mathrm{w}}$. If so, the turbulence model must then be able to function reliably in viscous flow regions.

Such models are usually described as "low Reynolds number". The expression does not refer to the Re of the flow based on the characteristic scales of the problem, *e.g.* axial mean flow speed and diameter for a pipe. Instead it is a "turbulence" Reynolds number $\mathrm{Re}_\tau$ based on the scales of speed $u$ and size $l$ of turbulent eddies and can be defined as

$$\mathrm{Re}_\tau = \frac{k^2}{\varepsilon\nu}. \tag{7.27}$$

This definition is obtained from scale arguments introduced in Sec. 6.6, in which $\varepsilon \sim u^3/l$. Since $u$ represents fluctuations, $k \sim u^2$. Combining these expressions into a Reynolds number yields Eq. (7.27).

## Asymptotic consistency

Low-$\mathrm{Re}_\tau$ turbulence models pay attention to the behaviour of fluctuating velocities, *e.g.* $u_x'$, $u_y'$, in the limit that $y \to 0$ at the solid boundary.

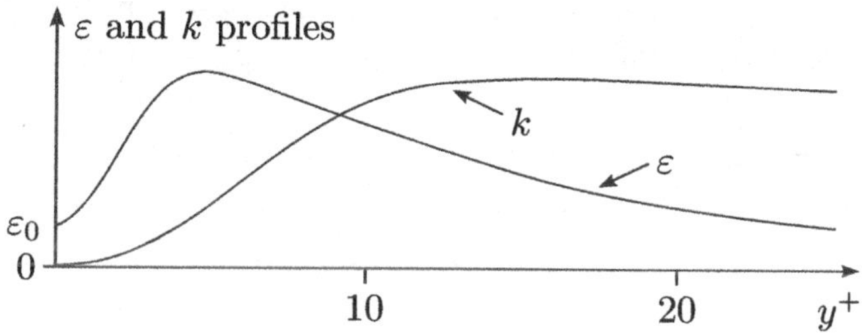


They aim to capture the shape the profiles of $\varepsilon$ and $k$ as they approach $y \to 0$. Let $x$ and $y$ define the directions tangential and normal to the wall respectively. Profiles in the fluctuating velocities can be expressed by polynomials in $y$, *i.e.*

$$u_x' = a_0 + a_1 y + a_2 y^2 + \dots$$
$$u_y' = b_0 + b_1 y + b_2 y^2 + \dots$$

where $a_0, b_0$, *etc.* are functions of space and time. The no slip condition implies $a_0 = b_0 = 0$, so $u_x' = a_1 y$ to the lowest order in $y$. For $u_y'$, it is $u_y' = b_2 y^2$ since, at the wall, $\partial u_x'/\partial x = 0$ and by continuity $\partial u_y'/\partial y = 0$.

The turbulent properties are, to the lowest order in $y$, as follows.

- From Sec. 6.11, $k = {}^1\!/_2\, \overline{u_x'^2} = {}^1\!/_2\, \overline{a_1^2} y^2 + \dots$
- From Eq. (6.30), $\varepsilon = \nu \overline{(\partial u_x'/\partial y)^2} = \nu \overline{a_1^2} + \dots$

It follows that models achieve asymptotic consistency when

$$k \sim y^2 \quad \text{and} \quad \varepsilon/k \to 2\nu/y^2, \tag{7.28}$$

in the limit that $y \to 0$.

## 7.9 Low-Re$_T$ $k$–$\varepsilon$ models

There are many low-Re$_T$ turbulence models for CFD simulations where the cells near the solid walls are sufficiently thin to resolve the flow through the viscous sub-layer.

Among them are several low-Re$_T$ $k$–$\varepsilon$ models based on Eq. (7.1) and Eq. (7.2) with additional corrections $f_1$, $f_2$, $\varepsilon_0$ and $E$:

$$\frac{\partial k}{\partial t} + \nabla \bullet (\mathbf{u}k) - \nabla \bullet (D_k \nabla k) = \\ G - \frac{2}{3}k(\nabla \bullet \mathbf{u}) - (\varepsilon + \varepsilon_0); \tag{7.29}$$

$$\frac{\partial \varepsilon}{\partial t} + \nabla \bullet (\mathbf{u}\varepsilon) - \nabla \bullet (D_\varepsilon \nabla \varepsilon) = \\ c_1 f_1 G \frac{\varepsilon}{k} - \frac{2}{3} c_1 f_1 \varepsilon (\nabla \bullet \mathbf{u}) - c_2 f_2 \frac{\varepsilon^2}{k} + E. \tag{7.30}$$

The calculation of $\nu_\mathrm{t}$ also includes a correction $f_\mu$:

$$\nu_\mathrm{t} = c_\mu f_\mu \frac{k^2}{\varepsilon}. \tag{7.31}$$

This model was first presented by **Jones and Launder** in their seminal $k$–$\varepsilon$ publication.† They proposed functions for $f_\mu$, $f_1$, $f_2$, $\varepsilon_0$ and $E$, as well as the coefficients $c_\mu$, $c_1$ and $c_2$.

**Launder and Sharma** subsequently presented‡ the model with a modified $f_\mu$ function and the more established coefficients listed in Eq. (7.3).

†William Jones and Brian Launder, *The prediction of laminarization with a two-equation model of turbulence*, 1972.

‡Brian Launder and B.I. Sharma, *Application of the energy-dissipation model of turbulence to the calculation of flow near a spinning disc*, 1974.

The resulting model became known as the Launder-Sharma $k$–$\varepsilon$ model.[†] It is arguably the most popular low-$\mathrm{Re}_\tau$ model today.

The first notable modification to the standard $k$–$\varepsilon$ model is $\varepsilon_0$ (sometimes denoted "$D$") in Eq. (7.29). It is the dissipation rate at the wall ($y = 0$), see figure, Sec. 7.7, calculated by

$$\varepsilon_0 = 2\nu(\nabla \sqrt{k})^2. \tag{7.32}$$

The term equates to $\varepsilon_0 = (\partial \sqrt{k}/\partial y)^2$ in the boundary layer which is consistent with Eq. (7.28). The benefit of redefining the dissipation rate as $\varepsilon + \varepsilon_0$ is that the boundary conditions at a wall for the Launder-Sharma model are the same for $\varepsilon$ and $k$:

- **fixed value** $\varepsilon = 0$;
- **fixed value** $k = 0$.

The next significant modification is the $f_\mu$ function

$$f_\mu = \exp\left[-3.4/(1 + \mathrm{Re}_\tau/50)^2\right]. \tag{7.33}$$

This modification recognises that $\nu_\mathrm{t} \propto l_\mathrm{m}^2$ so decreases $\nu_\mathrm{t}$ through the buffer and viscous sub-layer to the wall, consistent with the decrease in $l_\mathrm{m}$ according to the van Driest model Eq. (7.12).

The extra term $E$ in Eq. (7.30) is a follows, designed so that $k$ matches its recognised peak value within the buffer layer:

$$E = 2\nu\nu_\mathrm{t}\left|\nabla\nabla\mathbf{u}\right|^2. \tag{7.34}$$

Finally, $f_1$ and $f_2$ provide damping of the production and dissipation terms close to the wall in Eq. (7.30). The standard functions are $f_1 = 1$ (*i.e.* no damping), and

$$f_2 = 1 - 0.3\exp\left(-\mathrm{Re}_\tau^2\right), \tag{7.35}$$

which gives $f_2 = 0.7$ at the wall.

[†]although **Jones-Launder-Sharma** $k$–$\varepsilon$ **model** would seem more equitable.

## 7.10 Specific dissipation rate

The $k$–$\varepsilon$ model is one of a family of *two-equation* models for turbulence. With two equations, the models can represent each of the scales, $u_{\mathrm{m}}$ and $l_{\mathrm{m}}$, which characterise $\nu_{\mathrm{t}}$. Most often, $k$ is used to represent $u_{\mathrm{m}} \propto k^{1/2}$.

The other variable must represent $l_{\mathrm{m}}$ and so far we have used $\varepsilon \propto u_{\mathrm{m}}^3/l_{\mathrm{m}} = k^{3/2}/l_{\mathrm{m}}$ with SI units $\mathrm{m}^2\mathrm{s}^{-3}$. The *specific dissipation rate* $\omega \propto u_{\mathrm{m}}/l_{\mathrm{m}} = k^{1/2}/l_{\mathrm{m}}$, with SI units of $\mathrm{s}^{-1}$, is a popular alternative for this variable in turbulence modelling.

While Kolmogorov first proposed a two-equation $k$–$\omega$ model,[†] the $k$–$\omega$ models used in CFD originate from Wilcox.[‡] Here, "models" is plural since there are several versions of $k$–$\omega$ model with modifications and additions from its original form.

The **original $k$–$\omega$ model** is presented below (with some changes to the original variable names), assuming $\rho =$ constant for direct comparison with $k$–$\varepsilon$ in Sec. 7.1.

$$\frac{\partial k}{\partial t} + \nabla \bullet (\mathbf{u}k) - \nabla \bullet (D_k \nabla k) = G - \frac{2}{3}k(\nabla \bullet \mathbf{u}) - c_\mu \omega k \tag{7.36}$$

$$\frac{\partial \omega}{\partial t} + \nabla \bullet (\mathbf{u}\omega) - \nabla \bullet (D_\omega \nabla \omega) = \gamma G \frac{\omega}{k} - \frac{2}{3}\gamma\omega(\nabla \bullet \mathbf{u}) - \beta\omega^2 \tag{7.37}$$

[†] Andrey Nikolaevich Kolmogorov, *Equations of turbulent motion in an incompressible fluid*, first published in Russian in *Izv. Akad. Nauk SSSR* **6**, 1941.

[‡] David Wilcox, *Reassessment of the scale-determining equation for advanced turbulence models*, 1988.

The standard model coefficients are

$$\begin{aligned} &D_k = \nu + \alpha_k \nu_t, \quad D_\omega = \nu + \alpha_\omega \nu_t \\ &c_\mu = 0.09, \quad \beta = 3/40, \quad \gamma = 5/9, \\ &\alpha_k = 0.5, \quad \alpha_\omega = 0.5. \end{aligned} \tag{7.38}$$

Comparing dissipation terms in Eq. (7.1) and Eq. (7.36) gives the relation $\varepsilon = c_\mu \omega k$. Substituting in Eq. (6.31) leads to a simple relation for turbulent viscosity, given by

$$\nu_t = \frac{k}{\omega}. \tag{7.39}$$

Inlet and initial estimates for $\omega$ can be calculated by

$$\omega_{in} = c_\mu^{-1/4} \frac{k^{1/2}}{l_m}, \tag{7.40}$$

using $l_m$, in a manner similar to $\varepsilon_{in}$ in Eq. (7.4).

With wall functions, the *boundary condition* applied to $\omega$ set a **near-wall cell value** according to

$$\omega = \begin{cases} c_\mu^{-1/4} k_P^{1/2} / \kappa y_P & \text{for} \quad y_P^+ > y_{tr}^+; \\ 6\nu / \beta y_P^2 & \text{for} \quad y_P^+ \leq y_{tr}^+. \end{cases} \tag{7.41}$$

The expression for $y_P^+ \leq y_{tr}^+$ $(\omega = 6\nu/\beta y^2)$ is a solution to the following equation for the viscous sub-layer where diffusion and dissipation terms dominate in Eq. (7.37):

$$\nu \frac{\partial^2 \omega}{\partial y^2} - \beta \omega^2 = 0. \tag{7.42}$$

The equivalent dissipation terms for $\omega$ in Eq. (7.37) and $\varepsilon$ in Eq. (7.2) are $-\beta\omega^2$ and $-c_2\varepsilon^2/k$ respectively. The former is more stable in a numerical solution since it is insensitive to variations in $k$.

## 7.11 Enhancements to the $k$–$\omega$ model

A comparison of the $k$–$\omega$ and $k$–$\varepsilon$ models shows the dissipation term is $-\beta\omega^2$ in Eq. (7.37) for $\omega$ and $-c_2\,\varepsilon^2/k$ in Eq. (7.2) for $\varepsilon$. While the $\omega$ term behaves well, the $\varepsilon$ term is troublesome at a wall as $k \rightarrow 0$, so requires damping in the low-$\mathrm{Re}_T$ formulation in Sec. 7.9 to resolve the viscous sub-layer.

Its better dissipation term gives the $k$–$\omega$ model an advantage over $k$–$\varepsilon$ in the near-wall region. The disadvantage the original $k$–$\omega$ model is its sensitivity to the freestream values of $\omega$, which is not present in $\varepsilon$ in the $k$–$\varepsilon$ model. Neither model, in their original form, performs well under adverse pressure gradients.

However, since its initial publication, many enhancements have been made to the original $k$–$\omega$ model, in particular to address the problems mentioned above.

### Cross-diffusion

The dependency on freestream values of $\omega$ is addressed by the inclusion of a *cross-diffusion* term $\propto \nabla k \bullet \nabla\omega$ in the $\omega$ equation.† The term is derived when the $\varepsilon$ equation, *e.g.* Eq. (7.2), is expressed in terms of $\omega$ by substituting $\varepsilon \propto \omega k$. Its form is due to the expansion of $\nabla^2(\omega k)$ in the diffusion term by Eq. (2.74b).

The cross-diffusion term makes the $\omega$ equation more equivalent to $\varepsilon$, and thus independent of freestream values.

### Stress limiter

The original $k$–$\omega$ and $k$–$\varepsilon$ models are known to delay or suppress flow separation under adverse pressure gradients (described in

†Florian Menter, *Zonal two equation $k$–$\omega$ models for aerodynamic flows*, 1993.

Sec. 6.5). Under such conditions the ratio of the production to dissipation of $k$ can be significantly higher than unity. The calculated $\nu_t$ from Eq. (7.39) is excessively high, causing an over-prediction of shear stress $\tau_{yx} = \rho\nu_t\, \partial u_x/\partial y$.

The problem is alleviated by limiting the shear stress, based on the assumption it is proportional to $k$ in the boundary layer, *i.e.* $\tau_{yx} = \rho a_1 k$, where $a_1$ is a constant. A stress limiter is implemented through a modification to the calculation of $\nu_t$:

$$\nu_t = \frac{k}{\max\left(\omega, \Omega/a_1\right)}, \tag{7.43}$$

where $\Omega = \sqrt{2\mathbf{D}\boldsymbol{:}\mathbf{D}}$ is a 3D representation of $\partial u_x/\partial y$.

## Standard models

Different versions of the $k$–$\omega$ model are well catalogued in the *Turbulence modeling resource*, NASA Langley Research Center, `https://turbmodels.larc.nasa.gov`.

Today, there are arguably two "standard" $k$–$\omega$ models, First, the **$k$-$\omega$ SST model,**† (SST = shear stress transport) which emerged as a popular choice in CFD over recent decades.

It combines the $k$–$\omega$ model near the wall with $k$–$\varepsilon$ (expressed in terms of $\omega$) in the freestream, by applying *blending functions* to model coefficients, the cross-diffusion term and the stress limiter.

Secondly, the **$k$-$\omega$ 2006 model**‡ applies the cross-diffusion term and stress limiting to the original $k$–$\omega$ model. The terms are applied using *switches* so that the model maintains its simplicity, without the need for blending functions.

†Florian Menter, *Two-equation eddy-viscosity turbulence models for engineering applications*, 1994.

‡David Wilcox, *Turbulence modeling for CFD, 3rd ed.*, 2006.

## 7.12 Heat transfer in turbulent flow

The initial focus of turbulence modelling is to capture the effect of mixing on momentum diffusion since it influences the overall flow solution. But other properties are also transported by the turbulent eddying motions, in particular heat.

The effects of turbulence on heat transfer can be described using the following equation for internal energy $e$, obtained by substituting the material derivative in Eq. (2.57) and ignoring $\rho r$:

$$\frac{\partial \rho e}{\partial t} + \nabla \cdot (\mathbf{u}\rho e) = -\nabla \cdot \mathbf{q} + \boldsymbol{\tau} : \nabla \mathbf{u} - p(\nabla \cdot \mathbf{u}). \tag{7.44}$$

In turbulent flow, internal energy can be decomposed into averaged and fluctuating components $e(t) = \overline{e} + e'(t)$, see Eq. (6.11). The ensemble average of the terms in $e$ introduces a heat flux

$$\mathbf{q}_{\mathrm{t}} = -\rho\overline{\mathbf{u}'e'}. \tag{7.45}$$

This additional heat flux in the energy equation is equivalent to the Reynolds stress $\boldsymbol{\sigma}^{\mathrm{t}}$, Eq. (6.14), in momentum. Boussinesq modelled $\boldsymbol{\sigma}^{\mathrm{t}}$ by Eq. (6.20), using the concept of an eddy viscosity $\nu_{\mathrm{t}}$ corresponding to turbulent mixing, analogous to viscosity due to molecular motion according to Newton's law Eq. (2.41).

Similarly, $\mathbf{q}_{\mathrm{t}}$ can be modelled using a turbulent thermal conductivity $\kappa_{\mathrm{t}}$ due to turbulent mixing, by analogy with Fourier's law Eq. (2.54) for conduction due to molecular interaction

$$\mathbf{q}_{\mathrm{t}} = -\kappa_{\mathrm{t}} \nabla T. \tag{7.46}$$

The total heat flux $\mathbf{q}$ in Eq. (7.44) is then expressed in terms of the combined turbulent mixing and molecular interaction, using an *effective thermal conductivity* $\kappa_{\mathrm{eff}} = \kappa + \kappa_{\mathrm{t}}$, as follows:

$$\mathbf{q} = -\kappa_{\mathrm{eff}} \nabla T. \tag{7.47}$$

## Modelling turbulent heat transfer

Turbulent heat transfer can be incorporated into turbulence models based on eddy-viscosity and Reynolds-averaging, with additional thermal wall functions.

First, the calculation of $\mathbf{q}$ by Eq. (7.47) requires $\kappa_t$ from the turbulence model. A common approach to calculate $\kappa_t$ is from $\nu_t$ based on an estimate of *turbulent Prandtl number*

$$\text{Pr}_t = \rho c_p \frac{\nu_t}{\kappa_t}. \tag{7.48}$$

$\text{Pr}_t \approx 0.7 - 1.0$ provides a good estimate for many fluids, with $\text{Pr}_t = 0.85$ often chosen as a default value for CFD calculations. For some more unusual fluids, *e.g.* liquid metals, $\text{Pr}_t > 1$.

## Wall heat flux

The calculation of heat transfer through boundary walls is an important aspect of a many CFD simulations. Near walls, the distribution of $T$ tends to mimic $u_x$.

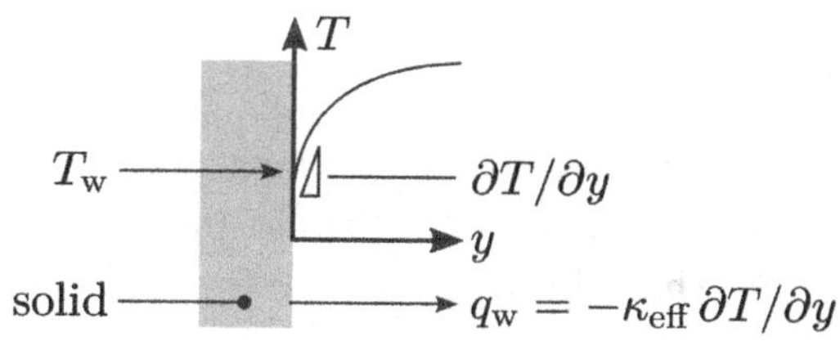


Consequently, the challenges of calculating wall heat flux $q_w$ are similar to wall shear stress $\tau_w$. Cells close to the wall must be very thin to resolve the viscous sub-layer in $T$ (when $\kappa_{\text{eff}} = \kappa$).

Otherwise, wall functions can be used to adjust $\kappa_t$ to compensate for the under-prediction of $\partial T/\partial y$ as described in Sec. 7.14.

## 7.13 Thermal boundary layers

In a turbulent boundary layer, the distribution of temperature is similar to velocity, with the viscous and intertial sub-layers, separated by a buffer layer, as discussed in Sec. 7.4.

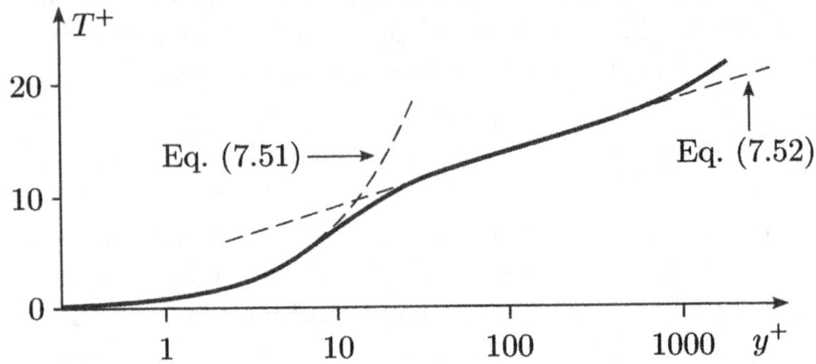


By analogy with $u_\tau$, Eq. (7.10), we define friction temperature as

$$T_\tau = -\frac{q_\mathrm{w}}{\mathrm{Pr}\,\kappa}\,\frac{y}{y^+}, \tag{7.49}$$

The wall layer is then described by a dimensionless temperature

$$T^+ = \frac{T - T_\mathrm{w}}{T_\tau}, \tag{7.50}$$

where $T_\mathrm{w}$ is the fluid temperature at the wall. Ignoring heat generation by viscous stresses, the profile in the viscous sub-layer is described by the relation

$$\boxed{T^+ = \mathrm{Pr}\,y^+} \tag{7.51}$$

The profile in the inertial sub-layer is commonly described by a log law for $T^+$

$$\boxed{T^+ = \frac{\mathrm{Pr_t}}{\kappa}\ln y^+ + B_T} \tag{7.52}$$

The derivation of Eq. (7.51) and Eq. (7.52) assumes a constant heat flux across the profile, equating to $q_{\mathrm{w}}$ at the wall. In the viscous sub-layer, the heat flux is laminar so $q_{\mathrm{w}} = -\kappa \partial T/\partial y$ and

$$T_\tau = \frac{1}{\mathrm{Pr}} \frac{\partial T}{\partial y^+}. \tag{7.53}$$

This equation integrates between $T$ at a distance $y^+$ from the wall to $T_{\mathrm{w}}$ at the wall, to yield Eq. (7.51). In the inertial layer, the heat flux is assumed turbulent $q_{\mathrm{w}} = -\kappa_{\mathrm{t}} \partial T/\partial y$ and

$$T_\tau = \frac{\nu_{\mathrm{t}}}{\nu} \frac{1}{\mathrm{Pr}_{\mathrm{t}}} \frac{\partial T}{\partial y^+}. \tag{7.54}$$

Combining Eq. (6.24), Eq. (7.9) and Eq. (7.15) yields the ratio $\nu_{\mathrm{t}}/\nu = \kappa y^+$. Substituting in Eq. (7.54) and integrating then leads to Eq. (7.52) where $B_T$ is the constant of integration.

The constant $B_T$ is generally considered to be a function of Pr. A reasonable approximation for this function is[†]

$$B_T = 13.7\,\mathrm{Pr}^{2/3} - 7.5. \tag{7.55}$$

Another function, commonly used in thermal wall functions, is $B_T = \mathrm{Pr}_{\mathrm{t}} B + P$, where $P$ is the function of $\mathrm{Pr}^* = \mathrm{Pr}/\mathrm{Pr}_{\mathrm{t}}$:[‡]

$$P = 9.24 \left[ (\mathrm{Pr}^*)^{3/4} - 1 \right] \left[ 1 + 0.28 \exp\left( -0.007\,\mathrm{Pr}^* \right) \right]. \tag{7.56}$$

The expression for $B_T$ uses the coefficient $B$ from Eq. (7.11). These constants of integration are sometimes subsumed within the log function as a coefficient "$E$" in the log law expressions, as the footnote on page 234 explains.

[†]Hermann Schlichting and Klaus Gersten, *Boundary-layer theory*, 2017.

[‡]Chandra Jayatilleke, *The influence of Prandtl number and surface roughness on the resistance of the laminar sub-layer to momentum and heat transfer*, 1966.

## 7.14 Thermal wall functions

Wall functions were introduced in Sec. 7.5 in order to improve the calculation of wall shear stress $\tau_w$ when cells are too large near a wall to resolve $\partial u_x/\partial y$ accurately. The same problem exists with heat flux $q_w$ and an under-predicted $\partial T/\partial y$. As before, the universal character of the boundary layer can be exploited, this time to improve the calculation of $q_w$.

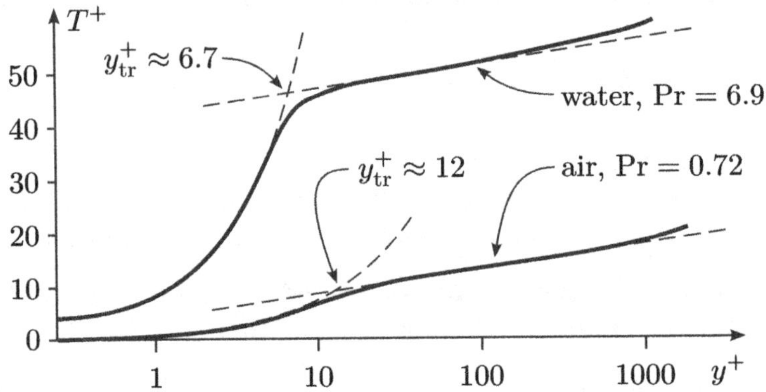


The temperature distribution is characterised by Eq. (7.51) for the viscous sub-layer, and the log law Eq. (7.52) for the inertial sub-layer. The transition $y^+_{tr}$ for $T$ occurs at the intersection of the two equations, *i.e.* when

$$y^+_{tr} = \frac{\mathrm{Pr_t}}{\mathrm{Pr}}\frac{1}{\kappa}\ln y^+_{tr} + \frac{1}{\mathrm{Pr}}B_T \tag{7.57}$$

While the transition between these regimes is fixed at $y^+_{tr} \approx 11$ by Eq. (7.18) for the $u_x$ profile, the (iterative) solution of Eq. (7.57) is dependent on Pr and $\mathrm{Pr_t}$.

Using $\mathrm{Pr_t} = 0.85$ and Eq. (7.55) for $B_T$, $y^+_{tr} \approx 12$ for air at 20 °C with Pr = 0.72. For water under the same conditions, Pr = 6.9 and the corresponding $y^+_{tr} \approx 6.7$.

A wall function can be derived which adjusts the turbulent conductivity $\kappa_\mathrm{t}$, in a similar manner to $\nu_\mathrm{t}$ in the standard wall function in Sec. 7.5. The model calculates $\kappa_\mathrm{t}$ for each patch face based on the near-wall cell $y_\mathrm{P}^+$.

No adjustment is made to $\kappa_\mathrm{t}$ when $y_\mathrm{P}^+$ corresponds to the viscous sub-layer. When $y_\mathrm{P}^+$ corresponds to the inertial sub-layer, $\kappa_\mathrm{t}$ is calculated as

$$\kappa_\mathrm{t} = \kappa_\ell \left( \frac{\mathrm{Pr}\, y_\mathrm{P}^+}{\mathrm{Pr_t} \ln y_\mathrm{P}^+ / \kappa + B_T} - 1 \right) \quad \text{for} \quad y_\mathrm{P}^+ > y_\mathrm{tr}^+, \tag{7.58}$$

where $\kappa_\ell$ denotes the laminar thermal conductivity, *i.e.* $\kappa$ from Eq. (2.54), to distinguish it from Kármán's constant $\kappa$. Eq. (7.58) uses $y_\mathrm{P}^+$ from Eq. (7.19), as in the standard $\nu_\mathrm{t}$ wall function.

The wall function is derived based on adjusting $\kappa_\mathrm{t}$ to improve the numerical calculation of $q_\mathrm{w}$ by

$$q_\mathrm{w} = -\kappa_\mathrm{eff} \frac{\partial T}{\partial y} = -(\kappa_\mathrm{t} + \kappa_\ell) \frac{T - T_\mathrm{w}}{y}, \tag{7.59}$$

where $T$ represents a value close to the wall, *e.g.* in a near-wall cell. By comparison, Eq. (7.49) and Eq. (7.50) combine to give

$$q_\mathrm{w} = -\frac{y^+}{T^+} \frac{T - T_\mathrm{w}}{y} \mathrm{Pr}\, \kappa_\ell. \tag{7.60}$$

The heat flux $q_\mathrm{w}$ is consistent between Eq. (7.59) and Eq. (7.60) when

$$T^+ = \frac{\kappa_\ell}{\kappa_\mathrm{t} + \kappa_\ell} \mathrm{Pr}\, y^+. \tag{7.61}$$

When the log law Eq. (7.52) is then substituted in Eq. (7.61), it provides the thermal wall function model which adjusts $\kappa_\mathrm{t}$ according to Eq. (7.58).

## 7.15 Summary of turbulence modelling

- The $k$–$\varepsilon$ model solves transport equations for turbulent kinetic energy $k$ and dissipation rate $\varepsilon$, Sec. 7.1.
- It is the original of a family of two-equation models, which ultimately provide $\nu_t$ to calculate the turbulent stresses.
- Initial and inlet values for $k$ and $\varepsilon$ must be specified, which can be calculated from turbulent intensity $I$ and mixing length scale $l_m$, respectively, Sec. 7.2.
- $I$ and $l_m$ are often estimated using functions that fit experimental data for fully developed turbulent flow, Sec. 7.3.
- The $k$–$\omega$ models replace $\varepsilon$ by specific dissipation rate $\omega$, with equivalent expressions for initial and inlet values, Sec. 7.10.
- The two "standard" $k$–$\omega$ models today are the $k$–$\omega$ SST model and $k$–$\omega$ 2006 model, Sec. 7.11.
- Models can provide a turbulent conductivity $\kappa_t$ to calculate the turbulent heat flux for thermal problems, Sec. 7.12.

### Turbulent boundary layers

- Turbulent boundary layers include a thin viscous sub-layer adjacent to the boundary with a linear velocity profile and, further from the boundary, the inertial sub-layer with a log law profile, Sec. 7.4.
- Profiles in temperature are similar to those for velocity, with equivalent linear and log law relationships, Sec. 7.13.
- Very thin cells are generally needed to resolve the viscous sub-layer to calculate the velocity gradient $\partial u_x/\partial y$ at the wall accurately, Sec. 7.5.

- Such thin cells within the boundary layer region can increase the mesh to a size which is prohibitively costly to run.

## Wall functions

- Wall functions permit much larger cells near the wall, by exploiting the universal character of the velocity distribution.
- The functions increase $\nu_{\mathrm{t}}$ at the wall to compensate for the under-prediction of $\partial u_x/\partial y$ with larger cells, to improve the prediction of the wall shear stress, Sec. 7.5.
- Thermal wall functions similarly increase $\kappa_{\mathrm{t}}$ at the wall to compensate for the under-prediction of $\partial T/\partial y$, in order to improve the prediction of the wall heat flux, Sec. 7.14.
- Standard wall functions make no adjustment to $\nu_{\mathrm{t}}$ when the near wall cell centre falls below the transition within a buffer layer, Sec. 7.5.
- Other models include a continuous function of $\nu_{\mathrm{t}}$ through the viscous sub-layer to the wall and adjustments for surface roughness, Sec. 7.6.
- Boundary conditions for turbulence fields with wall functions are based on observed profiles of those fields, Sec. 7.7.

## Models with resolved boundary layers

- Turbulence models must predict the universal character of boundary layers when the viscous sub-layer is resolved with sufficiently thin cells, Sec. 7.8.
- Models like the Launder-Sharma $k$–$\varepsilon$ include source terms and damping functions to improve the predictions and to simplify boundary conditions, Sec. 7.9.

# Chapter 8

# Sample Problems

*"I know there's an answer, I know now but I have to find it by myself."*

The Beach Boys, *I Know There's An Answer* (1966).

The preceding chapters establish some general principles of CFD, covering topics in fluid dynamics, numerical methods, boundary conditions, algorithms, solvers and turbulence and its modelling.

This chapter aims to bring together some of these principles using four sample CFD problems, some of which are mentioned in the earlier chapters. The four examples demonstrate a variety of flow types, including:

- two *internal* flows, inside a room and a Venturi tube, in which, beyond regions of significant inflow and outflow, the fluid is contained by solid boundaries;
- two *external* flows, around a vehicle and a cylinder, in which a considerable part of the domain is unbounded.

The flow in the room and around the vehicle are turbulent, so demonstrate turbulence modelling and wall functions from Chapter 7. The flow is laminar in the other examples.

The vehicle and Venturi tube cases are simulated using the steady-state algorithm in Sec. 5.12, whereas the room and cylinder cases use the transient algorithm in in Sec. 5.19.

## Mesh generation

The sample problems are supported by additional information on mesh generation. Strategies for meshing between the far-field and a body of interest in external flows are discussed in Sec. 8.1.

Meshes may include cells with high aspect ratio to increase the solution efficiency, particularly with external flows. Including these cells, however, increases the range of magnitudes of the matrix coefficients, which adversely affects solution convergence, as discussed in Sec. 8.2.

The generation of block-structured and unstructured meshes are discussed in Sec. 8.3 and Sec. 8.5 respectively. The block-structured approach is suited to problems whose boundaries are described by relatively simple geometric forms, *e.g.* cylinders, cuboids, *etc.*

The unstructured approach is designed to accommodate more complex boundary geometries, *e.g.* the vehicle in the external aerodynamics example. The geometries are typically defined by a computer-aided design (CAD) model.

## Boundary conditions

The examples demonstrate a variety of specialised boundary conditions. The cylinder and vehicle both use the freestream condition, described in Sec. 4.16 at the far-field boundary. The Venturi example uses a nonuniform inlet condition for $\mathbf{u}$, described in Sec. 8.7. Finally, the flow in a room uses a condition for $p$ to ensure correct boundary fluxes, as described in Sec. 5.11 and the external heat flux condition for $T$ from Sec. 4.17.

## 8.1 External flows

This chapter includes examples of flow around a cylinder and automotive aerodynamics. They are examples of *external* flows, which are characterised by a lack of solid boundaries to define the extent of some part, or all, of the solution domain.

The cost of a CFD simulation increases with the size of the computational mesh, so we cannot afford to have a solution domain which extends indefinitely in some direction. Boundaries must to be drawn somewhere, but which locations are optimal?

To calculated forces on a body accurately, the external boundaries need to positioned a surprisingly long distance from the body. For example, CFD calculations of forces on an aerofoil have used a domain length of 1000× the aerofoil chord length.

To understand why large distances are necessary, we must first recognise that the pressure is imposed at external boundaries according to the applied boundary conditions. Thy can include a fixed value outflow condition in Sec. 4.3, a total pressure condition in Sec. 4.7, a freestream condition in Sec. 4.16, *etc.*

The pressure at external boundaries will inevitably differ from the pressure at the same location in an unbounded problem. As discussed in Sec. 2.22, a change in pressure at any point influences the pressure everywhere in the solution domain, so any discrepancies in boundary pressures are reflected in the force calculations, including the shear forces.

Larger solution domains yield smaller discrepancies in pressure, producing solutions that better represent an unbounded domain. Accepting large domains are necessary, we need to limit the computational cost with an efficient meshing strategy that combines smaller cells to resolve higher gradients in complex, transient flow regions, with larger cells where the flow is more uniform.

## Mesh refinement

An external flow approaches far-field freestream conditions in regions around the external boundaries. Larger cells are effective in these regions since the flow is fairly uniform; smaller cells can then be reserved for more critical regions, close to, and in the wake of, solid bodies.

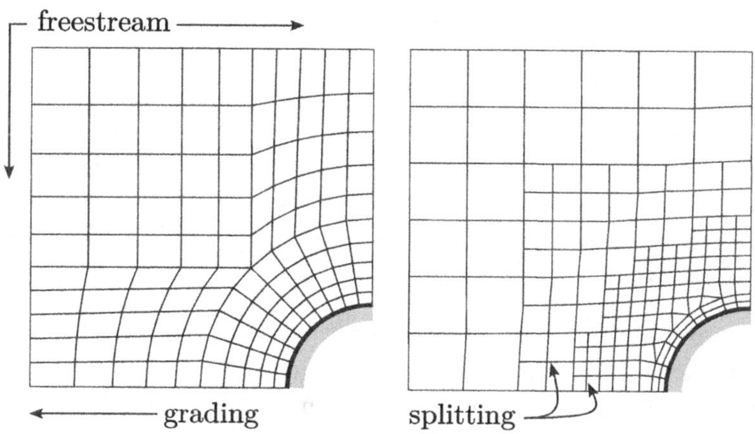


Two common ways to vary cell size are shown above. To the left is an example of a structured mesh which increases the cell lengths towards the freestream boundaries. The transition in size is smooth but cells tend to be generated which are noticeably longer in one direction than another, *i.e.* with a high *aspect ratio*, discussed in more detail in the following section.

To the right is an example of a mesh which refines the mesh away from the freestream boundaries using cell splitting. Starting from a coarse mesh, the method reduces the cell size by splitting each cell into two in each direction. This approach produces more abrupt changes in cell size across refinement interfaces, but maintains lower aspect ratios in the cells.

## 8.2 Aspect ratio

The previous section mentioned cells of a mesh that are longer in one direction than another. The parameter used to describe this phenomenon is *aspect ratio*, defined as the ratio of the longest to shortest dimensions, $l_{\mathrm{max}}/l_{\mathrm{min}}$, of a cell.

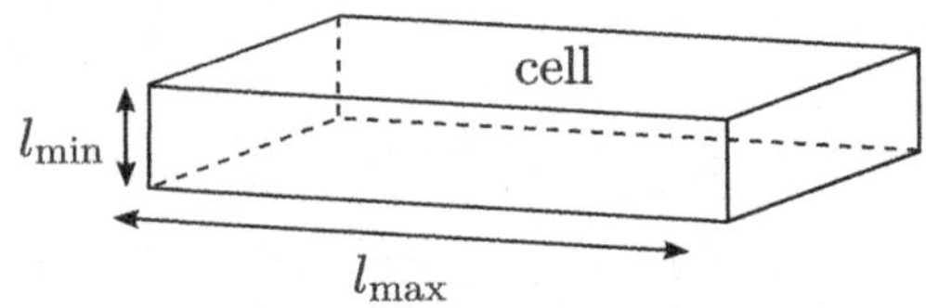


Meshes include cells with high aspect ratio for efficiency, by exploiting a prior understanding that the gradients in some flow properties *e.g.* shear stress, deformation rate and velocity, are much greater in one direction than others.

They are most commonly used to resolve the boundary layers at walls. Greater mesh resolution is required normal to the wall, where gradients are particularly high, than tangential to the wall. The centre heights of cells adjacent to the boundary need to be very small, corresponding to $y^+ = 1$.

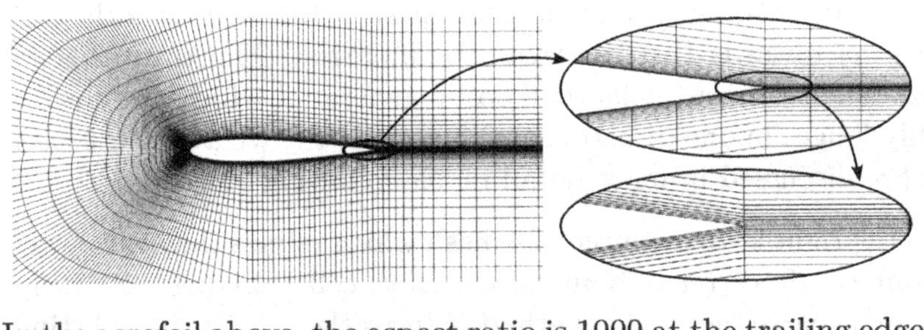

In the aerofoil above, the aspect ratio is 1000 at the trailing edge. If those cells were replaced by ones with an aspect ratio of 1, with the same cell height, the number of cells would increase by a factor of 1000, illustrating the extent of the gains in efficiency.

Introducing high aspect ratio in one area of the mesh causes problems elsewhere, however. The aerofoil example maintains a high aspect ratio beyond the trailing edge, actually increasing to $\sim 10^4$ further downstream. If the flow deviates from a direction parallel to the long side of the high aspect ratio cells, transient simulations then require very small $\Delta t$ to fall within reasonable Co limits. The mesh can be modified by abruptly reducing the aspect ratio beyond the trailing edge, but this increases the error associated with discretisation.

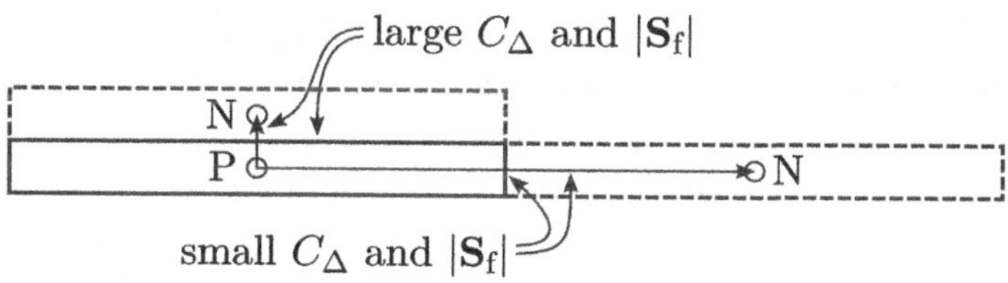


With Laplacian discretisation by Eq. (3.2) and Eq. (3.5), the matrix coefficients include a factor $|\mathbf{S}_\mathrm{f}|\, C_\Delta$. On a regular mesh, *both* $|\mathbf{S}_\mathrm{f}|$ and $C_\Delta$ are larger for the faces across the longest side of a high aspect ratio cell, than the shortest side, by a factor equal to the aspect ratio. Matrix coefficients can thereby differ by a factor of (aspect ratio)$^2$.

The convergence of the pressure equation, dominated by the Laplacian derivative, is adversely affected, in particular with the CG-based matrix solvers. The GAMG solver fares much better because the agglomeration strategy is based on a progressive reduction in aspect ratio as discussed in Sec. 5.18.

A matrix with a wide range of coefficient values yields lower residual values $r$, calculated from Eq. (5.11). In other words, for a given $r$, a solution with high aspect ratio cells is less converged than one without. Therefore, to maintain an equivalent level of convergence with high aspect ratio cells, tolerance controls need to be reduced, often by a few orders of magnitude.

## 8.3 Block-structured meshes

This section describes the generation of meshes for cases with boundaries formed by relatively simple geometric shapes, *e.g.* a cylinder, such as the examples in Sec. 8.4 and Sec. 8.8.

Block-structured mesh generation involves splitting the domain (assumed 3D) into several 6-sided blocks. Each block is then divided into a number of hexahedral-shaped cells.

There are some popular strategies for specifying the structure of blocks around boundaries of particular shapes. The most general approach uses blocks which extend outwards from sections of the boundary into the domain. For a cylinder in cross-section, 4 blocks are typically used as shown below by the blocks whose edges are oriented at 45° to the horizontal.

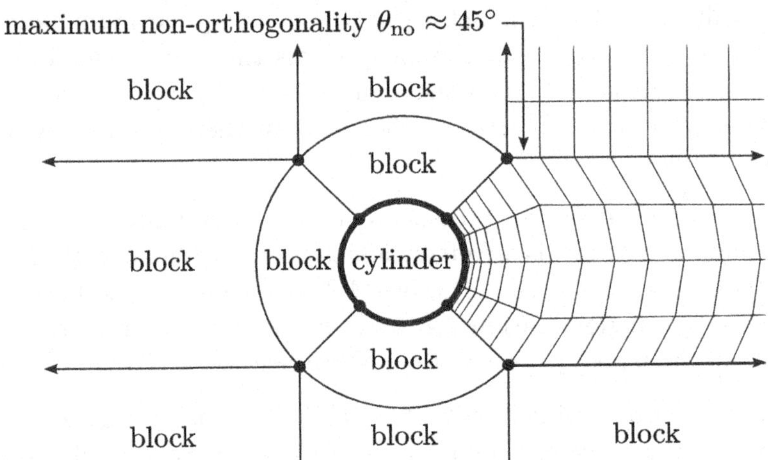


The domain extends away from the cylinder with additional blocks connected to those that encircle the cylinder. The mesh is usually graded starting with small cell heights on the cylinder surface, which gradually increase with distance from the surface. The main weakness of this block structure is the abrupt transition in non-orthogonality from $\theta_{\text{no}} = 0°$ to $\theta_{\text{no}} \approx 45°$ at the outer vertices of the inner blocks, indicated in the figure.

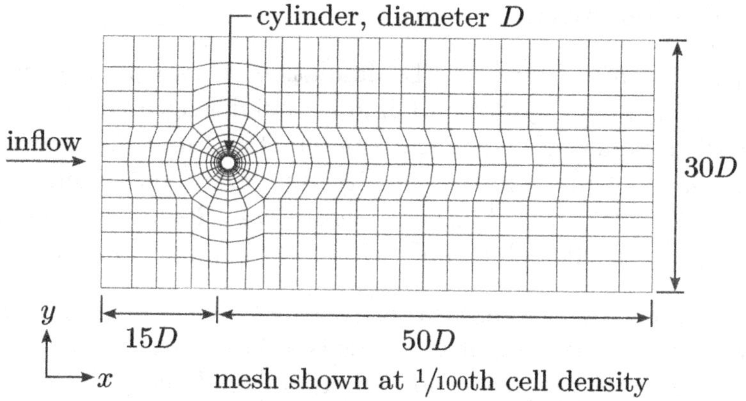


The figure above shows the structure of the mesh of 40,000 cells used in Sec. 8.4 for a cylinder of diameter $D$. Each element represents $10 \times 10 = 100$ cells in the actual mesh. There are 80 layers of cells around the cylinder, with grading that gives a centre height of $0.008D$ in cells adjacent to the cylinder surface.

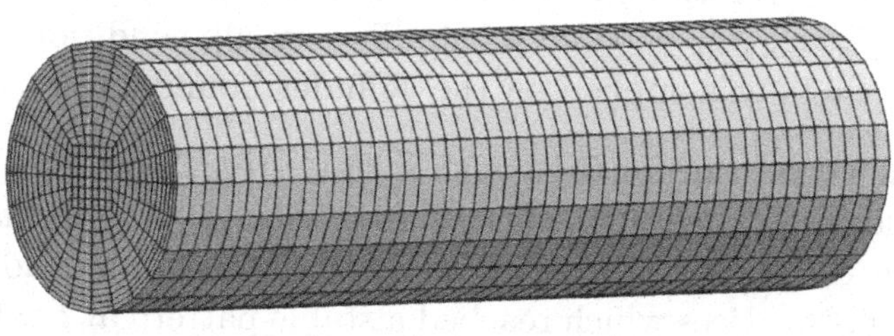

A mesh *inside* a cylinder, *e.g.* a pipe, can use a similar structure in which the inner vertices of the curved blocks are connected to form a single block along the centre, as shown above.

## 8.4 Flow around a cylinder

Flow around a cylinder was used as an example of boundary layer separation in Sec. 6.5. It showed an image of vortices shedding from the cylinder in a periodic manner, known as the Kármán vortex street.

The image comes from a CFD simulation in two dimensions, representative of a cylinder of infinite axial length, using the principal parameters: cylinder diameter $D = 2$ mm; freestream velocity $u_\infty = 1\ \mathrm{ms}^{-1}$ in the $+x$-direction; and, fluid kinematic viscosity $\nu = 10^{-5}\ \mathrm{m}^2\mathrm{s}^{-1}$. The corresponding Reynolds number $\mathrm{Re} = u_\infty D/\nu = 200$, which falls within the laminar flow regime.

The computational mesh used in the simulation is described in Sec. 8.3. The centre height of the cells adjacent to the cylinder was $0.008D = 16\ \mu\mathrm{m}$, corresponding to a calculated $y^+ < 1$.

The simulation used the transient solution algorithm in Sec. 5.19, solving for momentum conservation for an incompressible fluid, with $\mathbf{b} = 0$ and constant $\nu$. No energy equation was solved and no turbulence modelling was required since the flow was laminar.

The freestream boundary conditions from Sec. 4.16 were applied to $\mathbf{u}$ and $p$ over the entire external boundary, with reference values $u_x = u_\infty$, $u_y = 0$ and $p = 0$. The no-slip condition, $\mathbf{u} = \mathbf{0}$ and $\nabla_\mathbf{n} p = 0$, was applied on the cylinder boundary.

The simulation ran for 0.4 s with a time step $\Delta t = 25\ \mu\mathrm{s}$ using recommended numerical schemes from Sec. 3.23. Oscillations began in the wake of the cylinder at $t \approx 0.08$ s, soon leading to shedding of vortices which reached a stable pattern at $t \approx 0.15$ s.

The following figure, shaded by $|\mathbf{u}|/u_\infty$, shows the slower-moving vortices as the darker structures which propagate downstream of the cylinder.

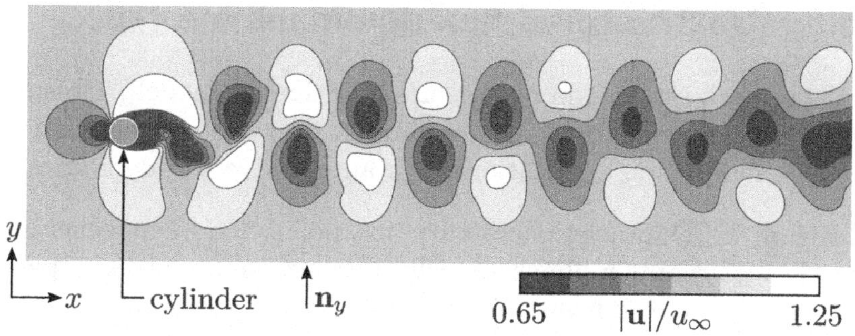


The repeated vortex shedding causes oscillations in the force $\mathbf{f}$ of the fluid *on* the cylinder. The force is calculated as the sum of viscous and pressure forces acting on the cylinder patch faces (f), in kinematic units $\mathrm{m^4 s^{-2}}$ (from kinematic $p$ and $\boldsymbol{\tau}$) by

$$\mathbf{f} = \sum_{\mathrm{f}} -\mathbf{S}_{\mathrm{f}} \cdot \boldsymbol{\sigma} = \sum_{\mathrm{f}} \mathbf{S}_{\mathrm{f}} \cdot (p\mathbf{I} - \boldsymbol{\tau}) = \sum_{\mathrm{f}} \mathbf{S}_{\mathrm{f}} p - \mathbf{S}_{\mathrm{f}} \cdot \boldsymbol{\tau}. \quad (8.1)$$

The dimensionless lift coefficient $C_{\mathrm{L}} = 2f_y/Au_\infty^2$ is shown below, calculated from: the $y$-component of the force is $f_y = \mathbf{n}_y \cdot \mathbf{f}$, where $\mathbf{n}_y$ is the unit vector in the $y$-direction; and, the projected frontal area $A$ of the cylinder.

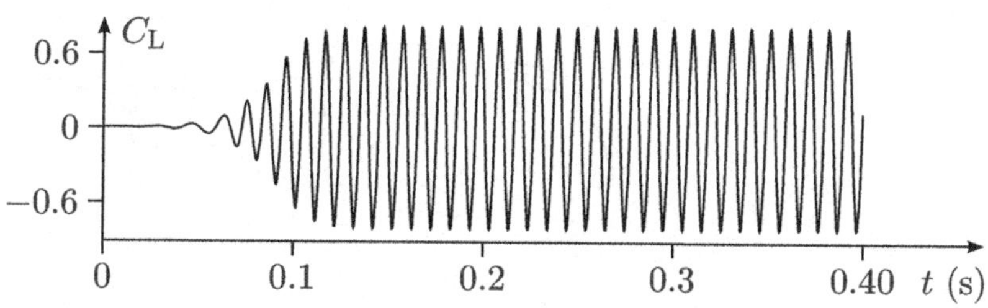


The oscillation period $\mathcal{T} = 0.0102\,\mathrm{s}$ corresponds to a Strouhal number $\mathrm{Sr} = D/(\mathcal{T} u_\infty) = 0.196$, consistent with published data.†

† Christoffer Norberg, *Flow around a circular cylinder: Aspects of fluctuating lift*, 2001.

## 8.5 Unstructured hex-dominant meshes

While block-structured meshes are effective for model problems with simple boundaries, *e.g.* a cylinder, they are impractical for many engineering applications, *e.g.* automotive aerodynamics.

Instead, CFD practitioners turn to tools which can generate meshes of many millions of cells and accommodate complex boundaries, *e.g.* a from a CAD model of a vehicle, in a highly automated manner.

A logical approach is to apply hexahedral-shaped cells that are oriented with the known, or anticipated, direction of flow. This is possible towards far-field boundaries and along solid walls, with the far-field cells generally larger than those at solid walls.

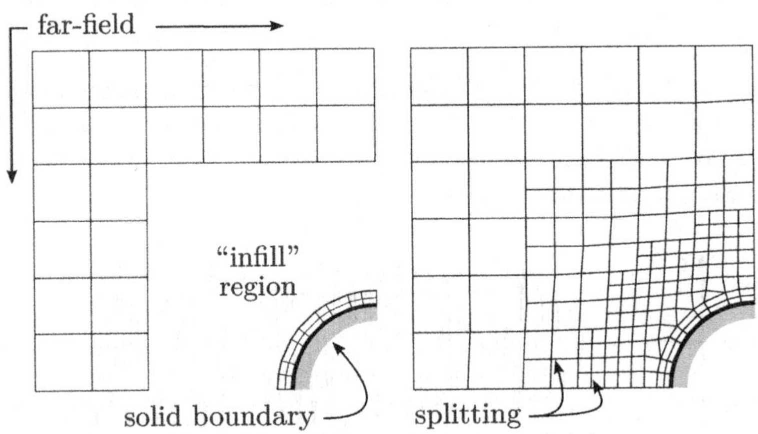


Cells must then be generated to fill the region between the aligned meshes near the far-field and solid boundaries. One approach is to infill the region with tetrahedral-shaped cells using automated methods based on Delaunay triangulation.[†]

[†]Boris Delaunay, *Sur la sphére vide*, 1934.

A more recent strategy is to fill the entire domain with large cells, similar to those in the far-field, which are then progressively split until they reach the desired size at solid boundaries. Subsequent steps then conform and align the cells to the solid boundaries.

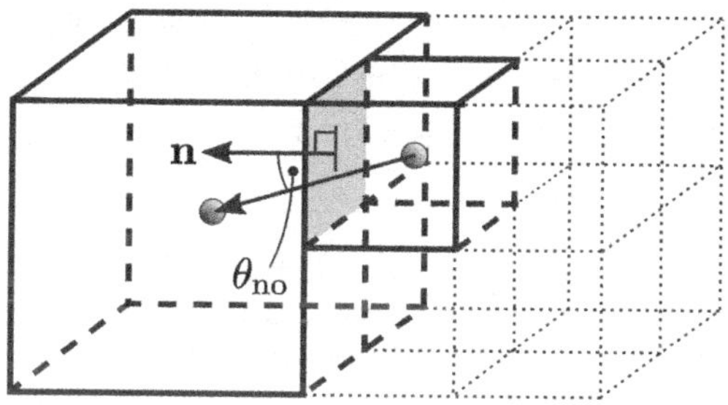


The infill region then contains mainly hexahedra, with the mesh described as "hex-dominant". Where there is transition in levels of cell splitting (see above), the larger cells become polyhedrons with an increased number of faces due to the smaller cells.

The splitting reduces solution accuracy by introducing a modest amount of non-orthogonality at the intersecting faces, *i.e.* $\theta_{no} \approx 25°$ for cells of unit aspect ratio. It also causes the weights $w$ for interpolation schemes, see Sec. 3.7, to deviate from the ideal $w = 1/2$; instead, for cells of unit aspect ratio, $w = 2/3$ (or $1/3$).

The generation of unstructured hex-dominant meshes is highly automatable, using simple controls that define the levels of cell splitting within specified regions of the domain, as shown above.

## 8.6 Automotive aerodynamics

An example of flow around a road vehicle was used to discuss some boundary conditions in Chapter 4 and to illustrate the cost of simulating turbulence in Sec. 6.8. An *aerodynamics* simulation was undertaken to capture the air flow around the vehicle, described by a CAD model. The aim was to calculate the drag coefficient at a speed of $u_\infty = 40\,\text{ms}^{-1}$.

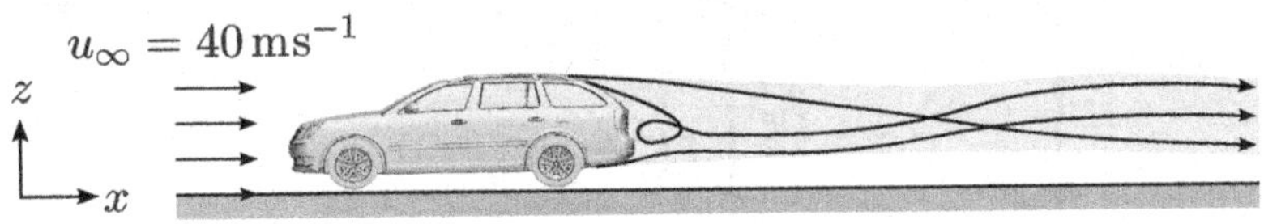


A mesh of 20 million cells was generated, with the vehicle facing a freestream flow velocity $u_x = u_\infty$. The vehicle and ground formed solid boundaries, with far-field boundaries positioned 10 m upstream and 30 m downstream of the vehicle.

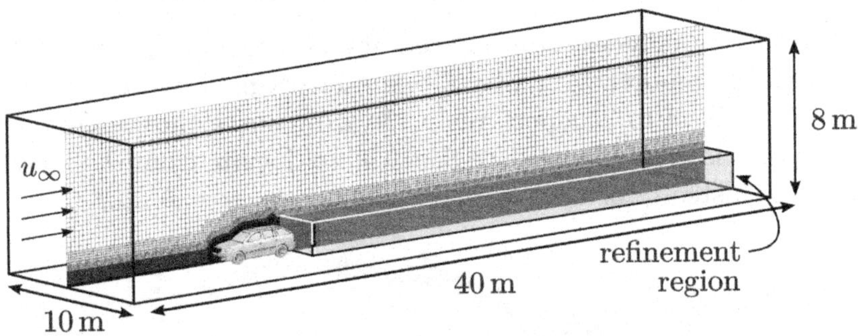


Along the elevated sections of the far-field boundary, the cell length was 0.25 m, reducing to 8 mm towards the vehicle by splitting within specified regions. Additional cell layers along the vehicle surface resulted in a near-wall cell height of 1.6 mm.

The simulation used the steady-state algorithm in Sec. 5.12, with an incompressible fluid with uniform $\nu = 1.5 \times 10^{-5}\,\text{m}^2\text{s}^{-1}$.

The freestream boundary conditions from Sec. 4.16 were applied to $\mathbf{u}$ and $p$ at the far-field boundaries, with reference values $u_x = u_\infty$, $u_y = 0$ and $p = 0$. The condition $\nabla_\mathbf{n} p = 0$ was a applied at solid boundaries, with $\mathbf{u} = \mathbf{0}$ applied to the vehicle and $u_x = u_\infty$ on the ground to emulate their relative motion.

Turbulence was modelled using the $k$–$\omega$ SST model described in Sec. 7.11. Turbulence levels of $k = 0.002\,\mathrm{m}^2\mathrm{s}^{-2}$ and $\omega = 1\,\mathrm{s}^{-1}$ were applied at the freestream boundaries and the standard wall function from Sec. 7.5 was applied at the vehicle and ground.

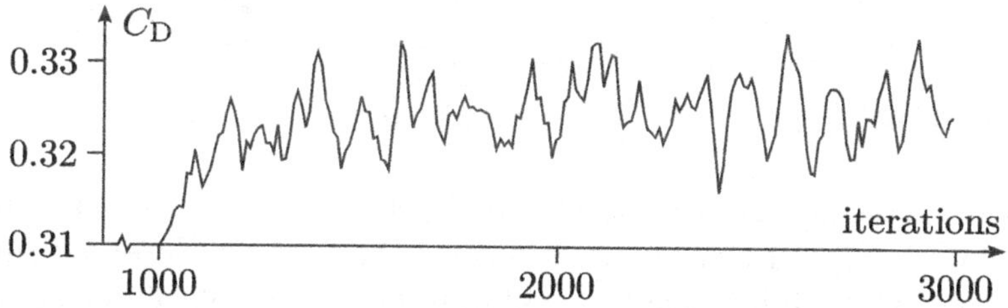


The simulation ran for 3000 iterations using numerical schemes recommended in Sec. 3.23. The drag coefficient $C_\mathrm{D} = 2 f_x / A u_\infty^2$ was calculated from the projected frontal area $A = 2.14$ m$^2$ and the $x$-component $f_x$ of the force $\mathbf{f}$ on the vehicle using Eq. (8.1).

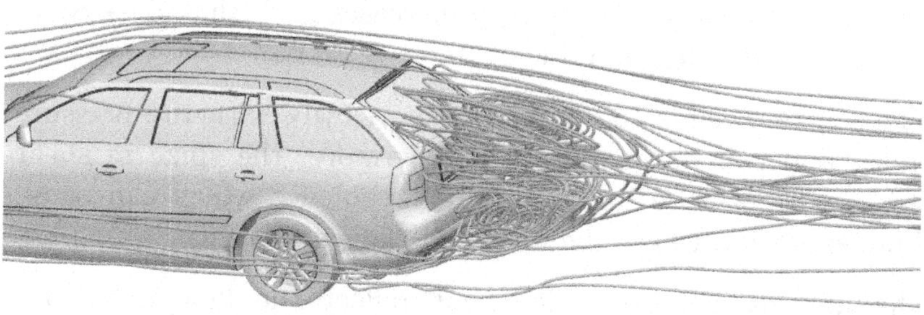

The flow in the wake of the vehicle is naturally unsteady, which prevents convergence to a steady-state solution. Beyond 1500 iterations, however, the solution oscillates around an estimated mean $C_\mathrm{D} \approx 0.325$.

## 8.7 Nonuniform inlet velocity

At an inlet boundary, the velocity is usually specified by a fixed value condition $\mathbf{u} = \mathbf{u}_{\text{in}}$, as discussed in Sec. 4.3. In many CFD applications, the flow at an inlet boundary is described by a single speed which is applied to all faces assuming $\mathbf{u}_{\text{in}}$ is uniform across the boundary.

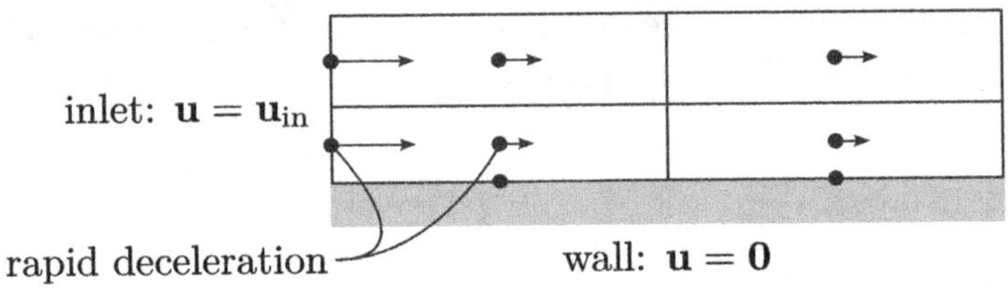


This creates an anomaly where the inlet boundary meets a wall. In the vicinity of the two boundaries, the flow decelerates from $\mathbf{u} = \mathbf{u}_{\text{in}}$ at an inlet face to a value in the adjacent cell close to the no-slip condition $\mathbf{u} = \mathbf{0}$ applied at the wall.

There is inevitably a high "spike" in pressure and shear stress within the cell in order to decelerate the flow so rapidly. As the length of the cell (in the flow direction) is reduced, the deceleration and associated pressure $p$ increases such that $p \to \infty$ in the limit that cell volume $V \to 0$.

The solution tends to converge more slowly with the pressure spike, and can be unstable. Furthermore, the spike in shear stress can generate high levels of turbulence which can cause flow separation where the inlet and wall meet.

The uniform condition does not reflect the flow behaviour upstream of the inlet. For example, assuming the wall extends upstream, a boundary layer would have developed at the inlet; or if the wall begins at the inlet, the flow would stagnate at its leading edge.

This is a good example of the axiom that numerical methods do not respond well to any modelling which is unphysical. The problems can be avoided by specifying a nonuniform $\mathbf{u}_{\text{in}}$ which represents the upstream flow better.

Some fields of engineering use established theories to describe the inlet $\mathbf{u}_{\text{in}}$, *e.g.* wind engineering uses a profile of $\mathbf{u}_{\text{in}}$ based on an atmospheric boundary layer along the earth's surface.

More generally, a nonuniform $\mathbf{u}_{\text{in}}$ can be specified which tends to $\mathbf{u} = \mathbf{0}$ at wall boundaries. Profiles can be described by $\hat{\mathbf{u}} = f(\hat{y})$ where: $\hat{\mathbf{u}} = \mathbf{u}/\mathbf{u}_{\text{max}}$ is the normalised velocity and $\hat{y} = y/y_{\text{max}}$ is normalised distance to the nearest wall boundary; and, $\mathbf{u}_{\text{max}}$ and $y_{\text{max}}$ denote the maximum $\mathbf{u}$ and $y$ values.

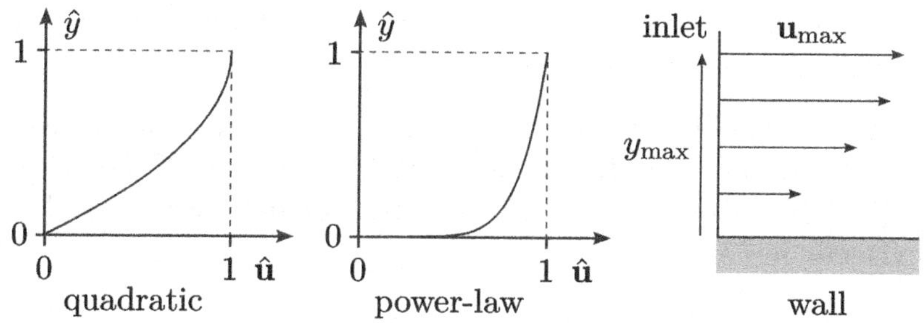


It is logical to specify $\mathbf{u}_{\text{in}}$ using established profiles for boundary layers as a reasonable estimate for the upstream conditions. A quadratic function $\hat{\mathbf{u}} = \hat{y}(2-\hat{y})$ represents a developed boundary layer for laminar flow, matching the analytical profiles for flow in a pipe or between flat parallel plates, *i.e.* Poiseuille's law.

Alternatively, a power law function $\hat{\mathbf{u}} = \hat{y}^{\,1/n}$ represents a developed turbulent boundary layer quite well. Prandtl used $n = 7$ — his *one-seventh power law*[†] — to reproduce data for flow in a pipe, but any suitable exponent $n$ can be used in practice.

[†]Ludwig Prandtl, *The mechanics of viscous fluids*, 1935.

## 8.8 Venturi tube

A Venturi tube is a device used to measure the flow rate of a fluid in a pipe. The device exploits the Venturi effect†, *i.e.* that pressure reduces when the fluid flows through a restriction.

The volumetric flow rate $Q$ is calculated according to

$$Q = \frac{C}{\sqrt{D/D_\mathrm{t} - 1}} A \sqrt{2\Delta p}, \tag{8.2}$$

where: $\Delta p$ is the decrease in *kinematic* pressure $p$ between the Venturi inlet and throat; $D$ and $D_\mathrm{t}$ are the inlet and throat diameter, respectively; the cross-sectional area is $A = \pi D^2/4$; and, $C$ is a *discharge coefficient*. A coefficient of $C = 1$ satisfies the Bernoulli equation, $u^2/2 + p =$ constant, between the inlet and throat, where $u$ is the fluid speed.

For $\mathrm{Re} > 10^5$, $Q$ is calculated accurately using $C$ values between 0.97 and 1,‡ but, for $\mathrm{Re} < 10^5$, suitable values of $C$ decrease significantly below 1 in order to account for pressure losses due to viscous forces.

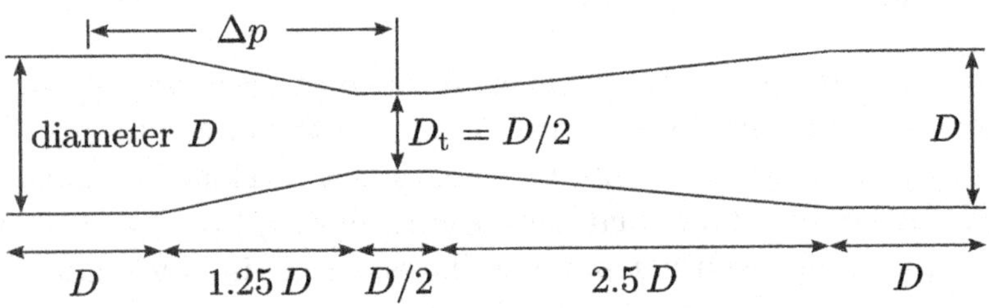


†Giovanni Battista Venturi, *Recherches expérimentales sur le principe de la communication latérale du mouvement dans les fluides appliqué a l'explication de différens phénomènes hydrauliques*, 1797.

‡ISO 5167-4, *Measurement of fluid flow by means of pressure differential devices inserted in circular cross-section conduits running full – part 4: venturi tubes*, 2003.

A simulation was performed to calculate $C$ for a Venturi tube, shown in the figure, with $D = 0.1\,\mathrm{m}$ and an inlet Re $= 500$. The inlet velocity was specified using the quadratic profile in Sec. 8.7 with a mean cross-sectional speed $u = Q/A = 0.2\,\mathrm{ms}^{-1}$, corresponding to a maximum speed $|\mathbf{u}_{\mathrm{max}}| = 0.4\,\mathrm{ms}^{-1}$.

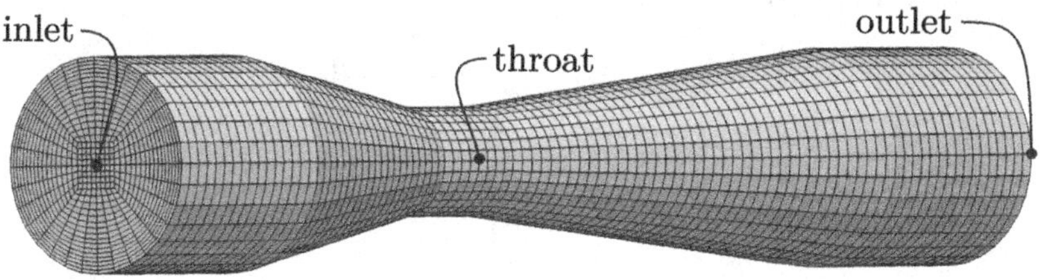


The simulation used the steady-state algorithm in Sec. 5.12, with an incompressible fluid with uniform $\nu = 4 \times 10^{-5}\ \mathrm{m}^2\mathrm{s}^{-1}$. The mesh contained 57,600 cells, with a near-wall cell height of 1.5 mm, which resolved the velocity profiles as shown below. The flow recirculates near the wall downstream of the Venturi throat, causing inflow at the outlet boundary. Consequently, the total pressure and inlet-outlet-velocity boundary conditions, described in Sec. 4.7 and Sec. 4.15 respectively, were applied at the outlet to maintain stability.

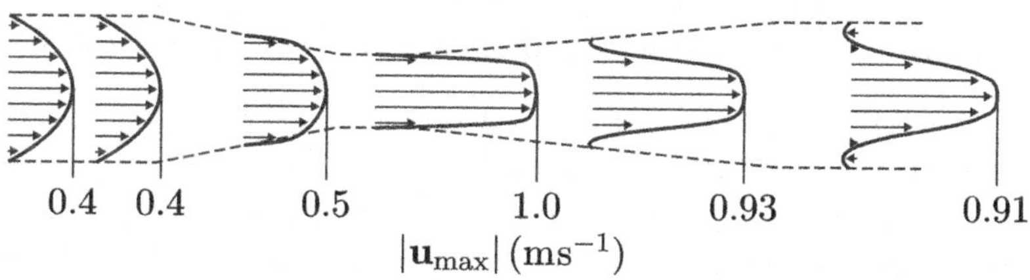


The flow was laminar so no turbulence modelling was required. The solution converged to within an absolute tolerance $r_{\mathrm{abs}} = 10^{-4}$, see Sec. 5.4, in 292 iterations.

The pressure drop was $\Delta p = 0.434\,\mathrm{m}^2\mathrm{s}^{-2}$ between the centres of the inlet and throat sections, with a corresponding $C = 0.83$.

## 8.9 Heating a room

A simulation of heating a room demonstrates natural convection driven by buoyancy forces. An idealised, ground floor room is presented below, with external glass doors and sloping roof, and internal walls and ceiling.

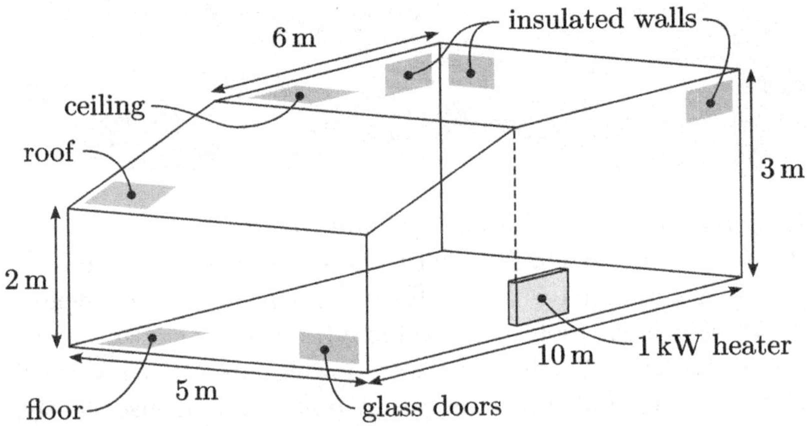


A heater is located along one side wall, below the point where the roof and ceiling meet. The aim of the simulation was to calculate the room temperature with the heater running at 1 kW, when the ambient external temperature $T_a = 270\,\text{K}$ ($\approx -3\,°\text{C}$).

The thermal boundary conditions were specified as follows: floor temperature $T_b = 285\,\text{K}$; ceiling $T_b = 294\,\text{K}$, representing the first floor temperature; insulated walls, with $q_n = 0$, see Sec. 4.17; glass doors and roof with a heat flux according to Eq. (4.30) using $h = 4\,\text{Wm}^{-2}\text{K}^{-1}$ and $h = 2\,\text{Wm}^{-2}\text{K}^{-1}$, respectively.

The mesh contained 350,000 hexahedral cells with grading that gave a cell height of approximately 20 mm along the walls.

Transport properties for air, $\nu = 1.5\times10^{-5}\,\text{m}^2\text{s}^{-1}$ and $\text{Pr} = 0.7$, were used. Turbulence was modelled using the $k$–$\omega$ SST model described in Sec. 7.11, with initial levels of $k = 0.001\ \mathrm{m^2s^{-2}}$ and $\omega = 1\ \mathrm{s^{-1}}$. The near-wall cell centres corresponded to $y^+ \approx 1-100$, so the continuous wall function from Sec. 7.6 and thermal wall function from Sec. 7.14 were applied at the boundaries.

The simulation used the transient solution algorithm in Sec. 5.19, including the buoyancy force $\mathbf{b} = (\rho/\rho_0)\mathbf{g}$ in Eq. (2.67), with $|\mathbf{g}| = 9.81\,\mathrm{ms^{-2}}$. The condition $\nabla_{\mathbf{n}} p = \mathbf{n} \bullet \mathbf{b}$ was applied at all boundaries, combined with the flux calculation in Eq. (5.20).

The variations in $\rho$ within $(\rho/\rho_0)\mathbf{g}$ were calculated using the ideal gas Eq. (2.55) using $R = 287\,\mathrm{Jkg^{-1}K^{-1}}$.

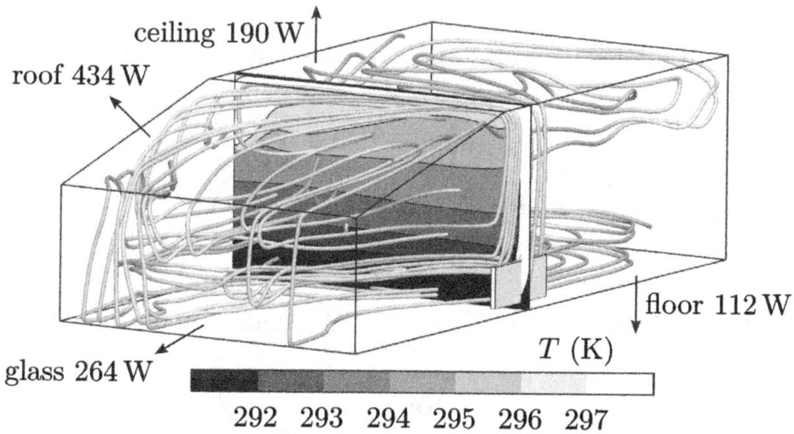


The simulation ran with a time step $\Delta t = 0.1\,\mathrm{s}$. The flow is highly unsteady, but at $t = 5000$ the heat losses through the boundaries oscillate about the mean levels indicated above.

Between the ground and 2 m, occupants experience a variation in $T = 291 - 295\,\mathrm{K}$ ($\approx 18 - 22\,°\mathrm{C}$). In the space adjacent to the roof and ceiling, the higher $T \approx 298\,\mathrm{K}$ generates significant heat losses, especially to the outside through the roof.

## 8.10 Building a CFD simulation

The examples in this chapter aim to demonstrate the general principles of CFD described in the previous chapters. They also highlight other important aspects of CFD, *e.g.* mesh generation. To finish with, here is some advice on building CFD simulations.

Before doing a simulation it is important to define its purpose, including the information you want to obtain. Initial data about the problem should be gathered, including the solution domain and boundaries, flow conditions, and models and properties that might be used. The flow Re should be estimated to establish whether the flow is laminar or turbulent (Chapter 6).

It is important to recognise that effective CFD simulations are not created in one attempt. Instead, they follow a typical design process, as shown below.

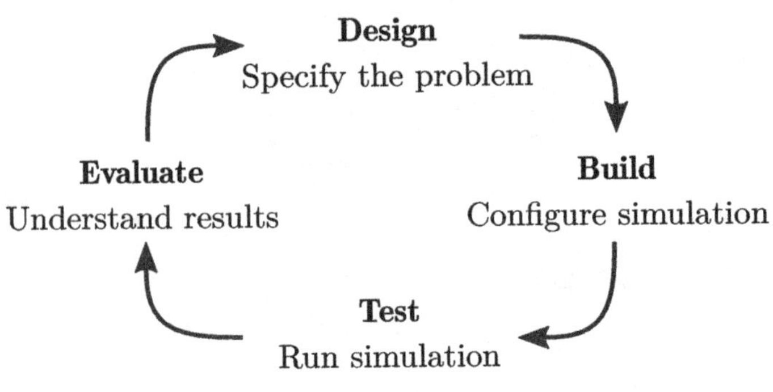


The process is iterative, beginning with a *prototype* simulation. The prototype should be as *simple* as possible in order to reach the first successful test quickly. Once achieved, the design cycle is set in motion and, thereafter, the simulation can be evaluated and improved in incremental steps, following the cycle above. Any problem can be attributed to the most recent change, which is much easier to diagnose when the change is small.

Simulations need to run quickly so that frequent, small changes can be tested efficiently. They run in a few minutes on a mesh of $\sim 10^5$ cells, which is a good initial size for the prototype. Steady-state solutions (Chapter 5) run particularly quickly.

Fields, *e.g.* $T$, must be initialised and the boundary conditions applied (Chapter 4) early in the design process. The prototype can include simple conditions, *e.g.* fixed value for $T$, before switching to more complex conditions, *e.g.* a heat flux for $T$.

If the flow is turbulent, robust RAS models should be deployed initially (Chapter 7). The prototype mesh size dictates that boundary layers are inevitably modelled using wall functions.

If a prototype simulation fails to run, numerical causes should be investigated, starting with mesh quality, *e.g.* non-orthogonality. Problems with discretisation schemes can be eliminated by applying the most stable schemes first, *e.g.* upwind (Chapter 3).

The initial simulations may simply establish a basic flow, solving mass and momentum conservation (Chapter 2) with simplifying assumptions, *e.g.* the incompressibility condition. The physical models should be simple, using constant properties.

The incremental changes to the simulation then incorporate additional models and equations. Often they require additional fields, boundary conditions, discretisation schemes, *etc.*

Beyond that, the simulation may venture into more complex areas of CFD modelling, including multiphase flows, conjugate heat transfer, compressible flows, reactions, particle methods and large-eddy simulation — outside the scope of this book.

However, the general principles presented in this book still apply whether a CFD simulation is simple or complex. And when simulations are more complex, they fail just as often because of a lack of adherence to these principles as for any other reason.

# Index

Made in the USA
Monee, IL
13 September 2026

62976816R00174